A Primer on Human Impacts on the Environment

A Primer on Human Impacts on the Environment

The Conceptual Approach

Liam Heneghan
Department of Environmental Science and Studies
DePaul University
Illinois, USA

Registered Offices
John Wiley & Sons, Inc., 111 River Street, Hoboken, NJ 07030, USA
John Wiley & Sons Ltd, The Atrium, Southern Gate, Chichester, West Sussex, PO19 8SQ, UK

Editorial Office
Boschstr. 12, 69469 Weinheim, Germany

For details of our global editorial offices, customer services, and more information about Wiley products visit us at www.wiley.com.

Library of Congress Cataloging-in-Publication Data
Names: Heneghan, Liam, author. | John Wiley & Sons, publisher.
Title: A primer on human impacts on the environment : the conceptual approach / Liam Heneghan.
Description: Hoboken, NJ : Wiley, 2023. | Includes bibliographical references and index.
Identifiers: LCCN 2022053743 (print) | LCCN 2022053744 (ebook) | ISBN 9781119642657 (paperback) | ISBN 9781119642718 (adobe pdf) | ISBN 9781119642619 (epub)
Subjects: LCSH: Nature–Effect of human beings on. | Global environmental change. | Environmental sciences.
Classification: LCC GF75 .H44 2023 (print) | LCC GF75 (ebook) | DDC 304.2/8–dc23/eng20230203
LC record available at https://lccn.loc.gov/2022053743
LC ebook record available at https://lccn.loc.gov/2022053744

Cover Design: Wiley
Cover Image: © Everett Collection/Shutterstock

Set in 9.5/12.5pt STIXTwoText by Straive, Pondicherry, India
SKY10043886_030223

To my granddaughter Iphigenia Rosemary Heneghan who was born as I wrote this book: may she and her generation confront the challenges of their times with an alacrity of spirit and flourishing hearts.

To the 900 students who, over the past two decades, have conceptualized alongside me the human engagement with our beautiful if battered world.

Contents

Preface: A Roadmap to the Book

It is time to change from a society based on conquest to a society based on survival.

Winona LaDuke 1992 [1]

In the autumn quarter of 2015, I instituted changes in a course that I had been teaching continuously since the beginning of the millennium. *Human Impacts on the Environment* was developed for college students majoring primarily in environmental disciplines. The course had initially taken a survey approach to the topic. In it we enumerated, as best as could be managed in a single 10-week quarter, all the major influences that humans – that most ubiquitous of primates – have had on life, atmosphere, water, rocks and soil, that is, upon the functioning of this planet Earth. After a decade and a half of pursuing an exhaustive approach to the material, five things had become apparent to me. Reflecting upon these resulted in a new presentation of the material.

This planet earth is the home upon which we rely for our sustenance, but it also serves as a sink for our wastes: both personal waste and the by-products of our economic activity. The fragility of the planet and of the human situation in the face of stresses of our own making has long been a subject of rumination in literature, the arts, and in scientific investigation. (Detail from The Garden of Earthly Delights (c.1500) by Hieronymus Bosch. Bosch's painting is from a triptych found in the Museo del Prado, Madrid, Spain). *Source:* Blankfaze / Wikimedia Commons / Public Domain.

As I will describe in more detail in the next chapter, the five considerations that refined my approach to thinking and teaching recognize that. . .

i) . . . what we know about the human impact on the biophysical processes of the planet is so extensive that taking a *conceptual approach* to teaching and thinking about global environmental problems is both an *efficient* way of presenting the material, but may be the only *feasible* way of learning about these issues in introductory courses;

ii) . . .in addition to what we already know about human impacts on the planet, new information concerning the damage to earth systems accumulates so rapidly that you, as a student, will benefit from practice in applying fundamental concepts to novel aspects of environmental change. The priorities that you will likely encounter during the course of your career may change, but many of the conceptual frameworks will endure;

iii) . . .the active role played by environmental scientists in policy and resource management decisions necessitates that you familiarize yourself with both the ecological and social context of environmental issues;

iv) . . .furthermore your responsibilities as a future collaborator obligate you to cultivate excellent interdisciplinary skills. The problems presented by global environmental problems are larger than any one discipline;

v) . . .because of a growing consensus that the consequences of environmental change are potentially *catastrophic* – and that these consequences become ever more calamitous if solutions are not prioritized – a student will need a grounding in understanding and applying risk assessment frameworks.

Making the necessary changes provoked by these fivefold interrelated factors not only transformed my teaching, but this reflection has also resulted in this book. Although the book you are reading has evolved largely out of my experiences of teaching environmental science over more than two decades, it is written, nonetheless, to serve the needs of students taking a *variety* of courses. This primer is designed for those studying the fundamentals of environmental science and studies. It should also be helpful as a supplement for courses focused on ecological restoration, green architecture and design, as well as on environmental management and urban planning. Finally, the book will provide useful background for students in the environmental humanities. The book aims at providing a versatile text for students learning about the environment in both formal and informal settings, as well as being suitable for early career environmental professionals, those who wish to become engaged in environmental work in an avocational way, and even for all those desiring to become well-informed global citizens.

1   What to Expect in this Book

In this book, we proceed as follows. In the first chapter of Section 1, the general philosophy adopted in this book – that is, placing the focus on *concepts* – will be justified (*Chapter 1: A Manifesto for Conceptual Thinking in Environmental Disciplines*). What, *precisely*, does it means to be conceptual in our thinking about the environment – how do conceptual models help frame the major problems, interrogate environmental mechanisms, and assist in envisioning solutions? These questions form the topics for *Chapter 2: A Conceptual Approach to Environmental Science*. Recognizing that much of the conceptual work that we plan to accomplish in this volume seeks to operationalize environmental concepts by defining them, the first section of the book concludes with *Chapter 3: A Short Chapter on the Definition of Definitions*.

This initial validation of a conceptual framing for the discipline is followed by a section presenting an overview of our basic models. Section 2 introduces the uniquely versatile and interdisciplinary tool of systems modeling. A variety of such models are applied in diverse fields (from economics, computer science, the social sciences to engineering, and, of course, they are widely used in environmental disciplines). We will introduce the general terms of these models in *Chapter 4: Everything is Connected: the First Rule of Ecology*. In this chapter we also evaluate the supposed central axiom of ecology that "everything is connected" and amend it. The specific details of the more significant environmental systems models are then assessed in *Chapter 5: Complex Environmental Systems*.

Although the term "environment," as we shall see, has been criticized as vague and hard to operationalize – how can or should this term be used? – in Section 3 we examine a sequence of increasingly sophisticated models that make something more concrete of this concept. *Chapter 6: All or Nothing? Or, What, Exactly, is an Environment?* introduces the basic concept of the environment. *Chapter 7: Life and Environment: the Indissoluble Link* illustrates how no organism can adequately be understood without consideration of how it is shaped by an environment, as well as, simultaneously, how it reciprocally shapes an environment. A final chapter in this section (*Chapter 8: Gaia, the Noösphere and the Anthropocene*) presents one very significant contemporary environmental model – that is, the notion of the Anthropocene (a term used to describe the most recent period in Earth's history when humans have had a marked impact on planetary functioning). The chapter also discusses how this model is both related to and different from other similar-seeming environmental models. These relatives of the concept of the "Anthropocene" are sometimes brushed aside as having little contemporary relevance; however, their utility may not have been fully exhausted. I try to show how they remain useful; we return to this theme in the final chapter of the book.

The idea of a "limit" is an especially important element of many environmental models. Section 4 of this book explores the concept of limits over the course of two chapters. The first of these chapters scrutinizes the consistent invocation of the notion of "limits" in environmental models (*Chapter 9: The Anthropocene and the Concept of Limits*). Positing a limit implies that in a finite world both population and economic growth must be bounded in some manner. Infinite growth, from this perspective, is not possible in a materially closed system. Several models propose that our contemporary global environmental problems emerge from our collective human enterprise now bumping up against the biophysical limits of the planet. This is the topic of *Chapter 10: Modeling the Limits*.

Will the breaching of environmental limits be catastrophic for humans and for the rest of nature? This is the topic of Section 5, which is the longest section of the book. The first of the four chapters in this section (*Chapter 11: Collapse and the Anthropocene: Learning from the Past*) examines the scholarship on the historical collapse of civilization(s). If there is an environmental component to the collapse of ancient civilizations, can this provide clues to our potential social vulnerabilities in the Anthropocene? *Chapter 12: How to Conceive of a (Climate) Crisis* illustrates the relevance of catastrophic risk assessment models for evaluating the potential consequences of global environmental problems. This chapter illustrates how risk models can be helpful in assessing problems associated with climate change. Risk models (both catastrophic and noncatastrophic) are regularly invoked in the work of The Intergovernmental Panel on Climate Change (IPCC), that is, the international body convened under the United Nations that collates scientific information on the causes and consequences of climate change, deliberates upon the significance of the results, and promotes international cooperation on climate action. After discussing climate risk, we evaluate risk associated with elevated species extinction rates, and the rapid decline in the ecosystem health of many key habitats on Earth – combined, these threats represent the potential unraveling of the fabric of life. The first of two chapters on biodiversity risk (*Chapter 13: Risking Life: Basic Concepts of Biological Diversity*) provides the necessary primer on biodiversity science; the second chapter (*Chapter 14: Is the Anthropocene Extinction a Global Catastrophe?*) applies specific risk models to contemporary extinctions rates.

It is, I think, worth pointing out here why considerable attention is devoted to biodiversity loss in this section of the book. There are two reasons. Firstly, more than ever before climate change and extinction risk are seen as deeply interrelated problems that need to be addressed together: I want to justify the immanent convergence of these agendas in these two chapters, as this decussation will influence that way in which we manage global environmental problems in the future. Secondly, we will take the opportunity in these chapters to provide many examples of how the conceptual approach adopted throughout the book can make complex problems more lucid than they might otherwise be. By the time you have reached these chapters you will be in a position to verify the strength of the conceptual approach, and will have developed some skill in applying the appropriate models.

In the concluding section (Section 6) we turn our focus more squarely on to the future, and interrogate how the conceptual modes discussed allow us to envision possible futures on this stressed planet. *Chapter 15: Conceiving a Future: the Need for Interdisciplinarity* provides tools to think about the future, and illustrates how all environmental disciplines lend an analytical power to conceiving the future. *Chapter 16: The Three Futures* presents three scenarios for the future: one is grim, another is highly desirable, and third – the likely future – falls between these two. Nudging our social and environmental future toward their ideal outcomes will require transformative visions. I end the book illustrating how *you*, the reader of this book, can be that oddest of all creatures, a butterfly whose wing-beat results in profound and positive improvements to the conditions of the world.

2   Some Features of the Book

This book's aspiration is to emphasize foundational concepts rather than offer an exhaustive account of each facet of environmental change – that is, the book serves as a primer – the chapters that follow are therefore written as concisely as possible. That being said, the text contains extensive references to both contemporary scientific literature as well as interdisciplinary classic texts from the history of the discipline. To facilitate the reader taking a deeper dive into the research literature, *all references for each topic are provide at the end of each chapter* (rather than being compiled in a single bibliography at the end of book). The reader is encouraged to browse these reference lists and follow up by reading more extensively on each topic, using, for example, Google Scholar – an accessible search tool that indexes scholarly literature. Many (though alas not all) of the resources can be found in a freely available form: you should also seek assistance from public and institutional libraries in tracking down other papers. If in doubt: ask a librarian!

The book is designed for accessibility: it should be relatively easy to read even for the novice environmental thinker. However, I also have made an effort to incorporate the extensive technical vocabulary of the environmental disciplines into the text. To balance approachability of the text with technical correctness in terminology I have provided a suite of *explanatory boxes* with glossaries, definitions, simple clarification of key concepts, occasional questions to ponder, and various other interventions that may be helpful to the reader. These will appear on the pages when they are immediately relevant rather than being assembled at the end of the book in a technical glossary section. The book contains a wide variety of figures: often these directly amplify the text on the pages on which they occur (and will be alluded to in the text). At other times, the material for figures is chosen to extend the concepts under discussion; they provide biographical information on leading conceptual thinkers, or refer to books and ideas from adjacent disciplines that will help illuminate your broader understanding of a topic. Though the text has a presentational priority, you should see figures and images in these boxes as important in amplifying the themes on each page.

This primer emphasizes the insights of the environmental sciences (both the natural and relevant social aspects of these disciplines), but in recognition that environmental issues are to some extent the intellectual terrain of nearly all intellectual endeavors, I have incorporated the perspectives of the humanities when this seems helpful. References are made, when it seems useful to do so, to philosophy, literature, history, and even, occasionally, to elements of pop culture.

Most of the chapters of this primer are accompanied by *fine-art images*. It is possible to justify the inclusion of art by resorting to elevated claims about the similarities and contrast between art and natural objects, or perhaps by pointing to the role of esthetics in our intellectual lives (for an impressive introduction to art and the life of the mind see Ross, Stephen David, ed. [1994] *Art and its Significance: An Anthology of Aesthetic Theory*. SUNY Press.) However, I argue for the inclusion of these images on a simpler basis: they are chosen to provoke an emotion, stimulate a thought, or offer a caesura – that is, a pause for reflection. The accompanying captions often provides a distillation of the topic of the chapter, or offers a thought or question to ponder. Think of these as being rather like a window out of which you might peer onto the world beyond the words on a page. Indeed, perhaps after reflecting on the images and text, look up from the book and gaze out upon the world asking Tolstoy's question: "What is to be done?"

Reference

1 LaDuke, W., *From Resistance to Regeneration*, in *The Nonviolent Activist* 1992.

Acknowledgments

Writing books is the most sociable of all the lonely pursuits. When taking up the pen, the writer occupies the isolating center of a mandala but it's a center that radiates out from the desk towards an environing and very garrulous world. This book is based upon the ruminations of half a lifetime during most of which time I was out and about in the world and, as often as not, both working with, and talking to, people. The preceding pages reflect these influences to a tremendous degree.

It been my good fortune to work within an extraordinary community of scholars, thinkers, and teachers at DePaul University. It is an institution that privileges the classroom – that most sacred of secular spaces – and one that acknowledges the possibilities for intellectual growth that emerge when a seasoned teacher, a host of bright young minds, and a set of difficult ideas are brought together (for 10 weeks at a time!). This *Primer* is a product, first and foremost, of discussions in the classroom, where we grappled with texts and ideas and contemplated the human engagement with the natural world. I thank all of these students sincerely, especially the many hundreds who have taken my *Human Impacts* course. As they learned the foundations of their discipline, I have learned a lot from their responses to the material. It is fitting, I think, to acknowledge those students who served as early readers of the book, and who commented upon, or copyedited, individual chapters or, in some cases, the entire volume: thanks to Caley Koch, Angela Stenberg, Jade Aponte, and Lizette Arroyo.

There have been many colleagues across the university whose willingness to put up with an interloper pillaging insights from their disciplines is appreciated. I am particularly thankful for colleagues in the philosophy department, especially Sean Kirkland, Rick Lee, and Will McNeill for their willingness to act as a sounding board for my interpretation of texts with which they are more intimately familiar. As for my home discipline in ecology and in the environmental sciences, I am very appreciative of my colleagues in the College of Science and Health where I have felt very well supported. My friends and collaborators in the Department of Environmental Science and Studies have been steadfast intellectual comrades over the past 25 years. Their insights are reflected in every page.

The number of colleagues outside my university who have helped me over the years are too many to name without risking forgetting someone. Rather than thank them all individually I will buy each of you a pint at an appropriate moment (upon presentation of short summary of a relevant chapter). I will mention one colleague and friend however: Dr. David Wise, emeritus professor of ecology in University of Illinois, Chicago, has been my ideal interlocutor over the years; I thank him and wish him well in his very active retirement.

I appreciate the support and hard work of all those at Wiley publishing who played a role in getting this book finished and out the door. The writing started in earnest during lockdowns associated with the COVID-19 pandemic, which was an arduous time for all of us in ways great and small. I feel fortunate to have been supported by Wiley both on the content editing side (especially Rosie Hayden and Frank Weinreich), but also during the production editing phase (with thanks to Shiji Sreejish and all her team, including Sivasri Chandrasekaran).

It's possible that this book would have been written whether or not I spent a few hours a week editing in the taproom of Sketchbook Brewing, in Evanston, Illinois, but would the process have been as enjoyable? The answer is no. Many big decisions about the manuscript were made in a contemplative corner of that establishment. Thanks to the staff for the community that they have fostered there.

May you be as lucky as I am in having a family that humors your foibles; mine are a center of gravity. Vassia Heneghan, the love of my life, my axis mundi, and who, for considerably more than half of my life, has been my first thought in the morning and my last thought at the end of each day: thank you, my love. Each page of this book is, in its own way, a (dense and inscrutable) love letter to you. Our boys, Fiacha and Oisín, are shimmering constellations. My thanks to them for reading endless texts consisting solely of draft paragraphs from this book: I impressed that that have not muted me. Thanks also to Sarah Heneghan, my beloved daughter-in-law, for providing several illustrations for this book.

My parents, Mary and Paddy Heneghan, provided the sort of childhood that allows a naturalist to develop; that is to say, I was governed with a mixture of a loose rein, exposure to the wilds, and much forbearance. My father passed away while I was writing this book. I don't know that he would have loved it but he would have loved showing it off. He was a great naturalist, and a man about whom many great stories will continue to be told. My mother remains the lodestar in our lives. Her great love of books, her indefatigable energy and entrepreneurial esprit remains a joy.

Finally, my gratitude to my granddaughter Iphigenia. Though only three months old she is an inspiration: every day she wakes up knowing her limitations and strives with every fiber of her being to overcome them, or at the very least to adapt to them. I have great confidence that her generation will find ways of flourishing in our tarnished world, and, indeed, may find ways of repairing and sustaining it.

Section One

Introduction to Primer

1

A Manifesto for Conceptual Thinking in Environmental Disciplines

One of the penalties of an ecological education is that one lives alone in a world of wounds.

Aldo Leopold, 1949 [1]

While this book was in preparation, an urgent and cautionary report was published by the Intergovernmental Panel on Climate Change (IPCC), the United Nations body responsible for advancing knowledge on human-induced climate change. Upon the release of the more than 3000-page report – the most comprehensive of the IPCC's climate reports yet – Hans-Otto Pörtner, a marine ecophysiologist and the lead editor of the report, stated that "the scientific evidence is unequivocal: climate change is a threat to human wellbeing and the health of the planet. Any further delay in concerted global action will miss a brief and rapidly closing window to secure a liveable future" [2]. If you are reading this book after the year 2023, the window will be narrower still. This tone of urgency – a livable planet is at stake – and the recognition that further delay in climate change action may well be catastrophic, echo similar expressions of concern over several other global environmental challenges. For example, when in 2019, the Intergovernmental Science-Policy Platform on Biodiversity and Ecosystem Services (IPBES) – the international organization convened to connect the best available biological diversity science and policy action – released *its* comprehensive report (*Global Assessment Report on Biodiversity and Ecosystem Services*), the chair of IPBES, Sir Robert Watson – a distinguished environmental chemist – stated: "The health of ecosystems on which we and all other species depend is deteriorating more rapidly than ever. We are eroding the very foundations of our economies, livelihoods, food security, health and quality of life worldwide" [3, 4].

Humans have always had to endure bad weather: it has dictated plans, caused havoc, and even inspired contemplation and art. Long-term and irrevocable changes in the climate – the patterns of weather experienced over a long period – are even more difficult (and expensive) to adapt to.

The IPCC sixth assessment report published in 2022 warns of increased heatwaves, droughts, and floods that are already causing great harm: lives lost, damage to infrastructure, and devastation for ecological communities.

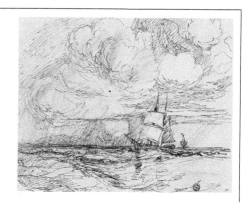

A Primer on Human Impacts on the Environment: The Conceptual Approach, First Edition. Liam Heneghan.
© 2023 John Wiley & Sons Ltd. Published 2023 by John Wiley & Sons Ltd.

Long-term changes in the climate will make extremes more pronounced. These changes will expose millions of people to food and water insecurity, and will often afflict those economic regions with the fewest resources to adapt to the change

Image: Ships at Sea during Storm (1830 / 49), Artist (assumed to be) Jules Dupré (French, 1811–1889), Art Institute of Chicago (designated as Public Domain).

There are several implications for the teaching of human impacts on the environment that can be gleaned from these reports as well as from the many thousands of scientific articles published on this topic each year. Global environmental problems are *complex* in their causes, *immense* in scope, *drastic* in potential consequences, and require large *interdisciplinary* networks of researchers and policy makers to address them. It should also be clear that global damage of this magnitude cannot be tackled independently of one another: loss of biodiversity, for example, is driven by a variety of factors, including climate change, and declining biodiversity can, in turn, have implications for *carbon sequestration*, which, in its turn, may exacerbate changes to the climate. Environmental challenges clearly cannot be tackled singly: if our response to the challenges needs to coordinated, then our *thinking about the environment* itself will need a set of coordinating principles – that is, they need *a strong conceptual foundation*. It is precisely this conceptual foundation – one that can facilitate clear thinking about distinct problems, but allows us to coordinate our responses at a global scale – that this primer seeks to provide.

Carbon sequestration refers to the transfer of some carbon dioxide (CO_2) out the atmosphere (where, as a greenhouse gas, it is a driver of climate change) and into other long-lived global carbon pools including those in the oceans, soil, geological strata, and living things. Increased sequestration in vegetation can therefore reduce the net rate of increase in atmospheric CO_2 (it is currently 418 parts per million [ppm] compared to pre-industrial levels of 278 ppm).

This introductory chapter sets out the case for a particular style of thinking (and teaching) about the environment that takes precisely this coordinated approach. In what follows I set out a case – let us call it a *manifesto* – for a conceptual education on environmental impacts.

A **manifesto** is a written document that forcefully proclaims a theory or cause, and often promotes a radical change in thinking or lifestyle.

1.1 The *Necessity* of a Conceptual Approach to Environmental Science

It has become quite evident that even if a course on human impacts on the environment is delivered over an entire academic year, or even two, there is simply too much information on the anthropogenic transformation of all aspects of earth systems to squeeze comfortably into such short period of time. After all, we have accumulated considerable information concerning those human impacts on the environment that have undoubtedly occurred since deep antiquity (see, for example, Charles Redman's book, *Human impact on ancient environments* published in 1999, which provides several case studies of human impacts that occurred in ancient times [5]) and there are allusions to the human modification of landscapes stretching back to the writings of the classical world. For example, references to ecological problems occur in the late Socratic dialogue *Critias*, where Critias and Timaeus (both of whose identities as historical figures are debated) are in conversation with Socrates (and with the Syracusan general Hermocrates making a cameo appearance); Critias remarked on the landscapes of Attica:

> But in the primitive state of the country, its mountains were high hills covered with soil, and the plains, as they are termed by us, of Phelleus were full of rich earth, and there was abundance of wood in the mountains. Of this last the traces still remain, for although some of the mountains now only afford sustenance to bees, not so very long ago there were still to be seen roofs of timber cut from trees growing there, which were of a size sufficient to cover the largest houses; and there were many other high trees, cultivated by man and bearing abundance of food for cattle. (*Critias* [6]),

Plato (approx. 424–348 BCE) was a philosopher of the classical age. Though not usually thought of as an environmental writer in the contemporary sense, nonetheless, like many writers of antiquity his writings, especially in the dialogues of Socrates can be mined for information about the changing landscapes of the ancient world. *Source:* [7] Nicholas Roerich 1893 / Public Domain

Despite the land's former excellence, Critias regrets the destruction of the Attic landscape over the course of his lifetime; these losses he attributes to complex causes, both natural and human. Although Plato's legacy as an ecologist is contentious, he is, without doubt, one of the earliest environmental writers in the Western tradition (though the anonymous author of the *Epic of Gilgamesh*, an epic poem from ancient Mesopotamia, written in the second millennium BCE – that is, more than a millennium before Plato – might also qualify as an early contributor to this tradition of thought [8].)

Setting aside such occasional mentions of the human stewardship of the landscape – or lack of it – in Plato, and more generally in the literature of antiquity, the first relatively modern treatment of the topic is by George Perkins Marsh (1801–1882). Marsh was an American polymath who published *Man and Nature, Physical Geography as Modified by Human Action* in 1864 [9]. Since that time, the field of environmental studies has burgeoned: environmental investigations have become highly technical, relying upon the use of sophisticated equipment to undertake sensitive measurements and employing vast computing and imaging power. By now, considerable amounts of information have been gleaned about most habitats on earth and from every region of the globe. As a result, the task of concentrating on even one small aspect of environmental change is challenging.

To illustrate: accumulating information about those aspects of the environment relevant for the research I conduct along with my students – that is, on biodiversity loss, the invasion of species into new geographical locations under human influence, and the restoration of degraded ecosystems – has been the work almost of a lifetime. It should be clear, then, that no single book (or university course) could possibly do justice to the immensity of the task of describing human impacts on the environment from antiquity to the present day. In response to the enormity of the undertaking, it seems preferable and more efficient to place a stress on the presentation of the *conceptual foundations* upon which *all* thinking about environmental change is based, rather than merely providing an exhaustive list of all the problems we are faced from ancient times until our present moment. These foundational concepts include, although are not restricted to, notions of environment, system's thinking, models of growth, rates of change, environmental thresholds and limits, disturbance, risk assessment, resilience, and a suite of concepts developed to capture our understanding of a planet transformed by human action. These latter conceptual models include the notion of the Anthropocene (and its predecessors) – the proposed term that denotes our present geological epoch where planetary metabolism is arguably regulated largely by human affairs and for human needs. The meaning of these conceptual terms will become apparent in the following chapters.

George Perkins Marsh, 1801–1882, a polymath with extensive knowledge of languages and other branches of knowledge wrote one of the earliest monographs in the English language on irreversible impacts of humans on the natural world. The volume: *Man and Nature, Physical Geography as Modified by Human Action* (1864). *Source:* Brian0918 / Wikipedia Commons / Public Domain.

In sacrificing aspects of a universal survey of the major environmental impacts now the core concepts are succinctly presented and illustrated along with a suite of relevant and detailed case studies. An overview of these core concepts can be accompanied by accounts of the oftentimes complex processes required to understand environmental impacts. In this way, the relevance of the core concepts are underscored by appropriate examples; the case studies, in their turn, reinforce an understanding of the concepts. Upon this conceptual foundation, students can develop a well-informed approach that reflects upon and informs action to address the major issue of our times, that is, the human transformation of the planet upon which we rely for resources, for ecosystem services, and as a source for inspiration and beauty.

By illustrating concepts with examples, students still engage with a multiplicity of facts, explanations of processes (all accompanied by the relevant vocabulary), but now instead of navigating a barrage of disconnecting facts, students will develop a unifying conception of how diverse problems are interrelated.

1.2   The *Necessity* of Adapting and Applying Standard Models to Novel Environmental Problems

A second, and related, motivation for the changes I introduced into my long-running course on human impacts – reflected in the chapters that follow – is in recognition of the accelerating pace with which damage to environmental systems is being discovered. Not only does the *pace of discovery* continue to accelerate, but the *rate of environmental change* under human influence is greater than ever before. That is, matters are getting worse faster. The last 50 years have seen the swiftest pace of anthropogenic change of earth systems in history [10]. The burgeoning scientific literature, along with both scientifically-informed journalism and increased public scholarship all have had the effect of creating a deepening sense of urgency among policy makers and the informed public. Increased public concern and policy attention has meant that research emphasis is placed upon key environmental problems (especially, though not exclusively, those related to climate change and biodiversity loss). This, in turn, has intensified the speed of further discovery in these research areas [11].

Environmental education must respond in a variety of ways to the accelerating pace of discovery. Not only do environmental professionals need comprehensive familiarity with the key conceptual frameworks of their discipline, they also need the acumen to apply generalized models flexibly to novel environmental situations. This is especially important as environmental change become more apparent. Research foci inevitably change as new aspects of environmental degradation come into view. In recent years, for example, we have witnessed an amplification of wildfire severity (often on vast scales) [12], an intensification of regional droughts [13], tropical storms that seemingly are becoming more intense, and undoubtedly more destructive [14, 15], greater levels of coral reef bleaching and destruction [16], an increase in marine acidification and eutrophication (nutrient pollution) [17], rapidly melting Antarctic ice [18], and so on. All of these examples serve to illustrate some of the emerging priorities for new investigation. Despite the rapid accumulation of this information, a resourceful environmental thinker should find themselves comfortably able to apply their conceptual training to these ever more complex situations. This is because new concepts generally evolve at a pace slower than the rapidly accumulating database of new instances of damage. The frameworks you glean from this book will be applicable in wider contexts.

To illustrate the idea that concepts evolve at a relatively stately pace, consider, for example, the continued relevance of **General Systems Theory**. This is an idea with ancient roots that was formalized in the middle of the twentieth Century, and which continues to provide, as we shall soon see, some of the primary conceptual tools for environmental analysis [19]. A systems thinker identifies the component parts of any given entity, evaluates the ways in which these components can interact, and assesses the manner in which the behavior of the whole emerges from multifaceted interactions among these parts. Systems analysis thus forms the basis for much hypothesis formation, experimentation, and data evaluation in the environmental disciplines [20]. Because the basic models of systems theory are formulated at a relatively low level of generality, a well-trained systems thinker can thus identify in newly emerging problems commonalties with previously familiar issues. A global structure may change and yet the student will be able to apply prior knowledge to these new situations in well considered and practical ways.

Despite the durable nature of many important conceptual models, such models can and, of course, do evolve. Ecosystem analysis, hierarchy theory, chaos theory, cybernetics, information theory, game and decision theory, resilience thinking, theories of combinogenesis, and most sustainability models, as we shall see, are merely elaborations of a more general systems theory.

1.3 The *Necessity* of Collaborative Approaches to Global Environmental Problems

As a result of recent cycles of discovery, growing public attention to an environmental problem, and a subsequent redoubling of focus on environmental research, students of the environment are no longer "living alone in a world of wounds." Contrary to the famous epigram quoted at the beginning of this chapter from wildlife ecologist and eco-philosopher Aldo Leopold (1887–1948), environmental students step out of their classroom and into a world in which environmental problems are frequently discussed. Deliberations over our environmental future are no longer confined to specialist communities of scientists and policy advocates, they are now vigorously discussed in magazines, newspapers, across social networks, in movies and popular shows, as well as in settings as various as church groups, public houses, and even around the family dinner table. Because of the greater levels of public awareness of environmental change, environmental professionals now have a responsibility for engaging with eclectic communities, often with varying levels of environmental literacy: this both presents opportunities and creates challenges for environmental scientists.

As a result of this, environmental practitioners often play an enhanced role in policy deliberations concerning our shared environmental future. Environmental scientists find themselves engaging in professional activities that are concerned with minimizing environmental risks to human health and well-being, as well as those contributing to sustaining economic prosperity. The distinction between disinterested science, applied research, and public action – which has traditionally been a vexed matter for the environmental disciplines (ecological scientists had often been encouraged to steer clear activism and politics [21]) – has been largely resolved in favor of environmental scholars playing prominent advisory roles in governance and management settings. Input from environmental scientists is necessary on scales from local communities all the way up to intergovernmental organizations.

A grounding in concepts is necessary when environmental specialists contribute to policy debates and governance processes. In addition to providing a way of framing environmental science, as we have already seen, conceptual models – often, though not always, based upon systems theory – have often been adapted for use in the environmental social sciences and in framing policy responses [22]. That is, conceptual tools are effective collaborative tools. It should not surprise us that forms of systems thinking provide a bridge between the natural sciences and social theory – endeavors that might formerly have seen themselves as distinct – since interdisciplinarity and public engagement had always been an aspiration of the architects of general systems theory [19]. For example, the theoretical biologist and systems theorist Ludwig von Bertalanffy (1901–1972) in his influential book, *General System Theory: Foundations, Development, Applications* (1968), calls quite explicitly for a unity of the heretofore disparate sciences [23]. Unsurprisingly, Von Bertalanffy's work explicitly identifies "air and water pollution" along with social problems such as urban blight, juvenile delinquency and organized crime as problems that call for systems' based solutions.,

Margaret Mead (1901–1978). Mead was a renowned anthropologist who in 1972 served as President of The International Society for the Systems Sciences (ISSS), which is one of the oldest interdisciplinary organizations devoted to understanding complex systems. Some of her systems' interests focused on the need for clearer communication among scientists, but also between scientists and the public. Mead wrote, "And we shall need still newer kinds of instrumentation – macroscopes that can simplify without distorting the complexity of our knowledge of the biosphere and the cosmos within which a recognition of all disciplined human endeavor must now take place" [24]. *Source:* Smithsonian Institution / Wikimedia Commons / Public Domain.

The work of anthropologist Margaret Mead (1901–1978), who served as President of *The International Society for the Systems Sciences* (ISSS) in 1972, emphasized the role of systems theory in collaboration. She stressed the significance of system theory in facilitating communication between scientists. The collaborative act of developing and applying systems models to a range of situations can in itself be understood and analyzed as a system process in its own right (this insight is sometimes referred to as the "cybernetics or cybernetics"). In this way, environmental students become part of the systems in which they participate when they begin to undertake the task of accurately depicting problems and are engaged in devising solutions. In doing so, however, they must ensure the consistency of policy formation with the best available science.

Although environmental education often focuses primarily on an academic training, students must, additionally, learn to be effective collaborators. This is especially important as policy makers, and often the general public (when policies are opened up for community comment), will often share a direct role in developing research questions and agendas [25, 26].

1.4   The *Necessity* of Evaluating Risk

Consistent with the ambition of most intellectual endeavors – the essence of which was first articulated by Aristotle (384–322 BCE) in his claim that "all human beings by nature desire to know" – environmental scientists, like other scientist, seek to provide "an organized account of whatever knowledge we can obtain about the universe" [27]. However, in addition to organizing our account of knowledge, science also helps us in evaluating future uncertainty. This application of knowledge to managing risk has become a central undertaking of the environmental sciences.

The *four subcomponents* of the earth system include the **atmosphere** (the gaseous envelope overlying the solid and liquid surface of a planet), **lithosphere** (the Earth's crust and a portion of the upper mantle), **hydrosphere** (the Earth's fresh and saline water): water can occur in a liquid, solid or gaseous state, and the **biosphere** (the sum of all of life forms on earth).

As their most basic task, the environmental disciplines pursue knowledge about earth systems. These sciences aim to provide a holistic account of how the four subcomponents of the earth system – atmosphere, lithosphere, hydrosphere, and biosphere – interact to produce the phenomena that collectively constitute that entity sometimes referred to as "Gaia" (that is, the co-evolving nexus of the living and nonliving elements that comprise the planet) [28]. By holistic is meant the tendency of many entities, natural ones in particular, to function as an integrated whole. More, however, than just organizing our knowledge about the natural world, the environmental sciences also try *to predict how earth systems may change in the future*. Predictions about the future state of the environment are made in order to evaluate the risks that humans and other living beings will face. Interrogations about the future are poised at the intersection of the environmental disciplines and the discipline of risk assessment.

Recognizing the potential social significance of environmental forecasting, as well as contemplating those changes that might have be taken to avert a catastrophe, requires a sophisticated approach to conceptual and mathematical modeling in environmental disciplines. To illustrate this complexity, consider, for example, those models that are often used to determine the future of climate, such as General Circulation Models (GCMs). GCMs are computationally intricate: they cast a conceptual grid over the planet with a horizontal spatial resolution of only a few hundred kilometers while also, on the vertical scale, evaluating as many as 30 slices of atmosphere over each spatial plot. Models of the behavior of these three-dimensional "pixels" are then combined in order to determine expected temperature changes for the entire earth. In general, predictions from environmental models, such as these GLMs, integrate detailed knowledge of the relevant environmental processes with information about the possible responses of a system to modifications in key driving variables (for example, how are global temperatures modified by the altered gaseous composition of the atmosphere or how might an ecosystem respond to the loss of critical species?). Outputs from the models are then extrapolated out from the short-term to many decades away in order to create plausible scenarios about the future [29].

Weather refers to short-term changes in atmospheric conditions: often measured in hours, days or weeks. In contrast, **climate** refers to weather conditions in a particular region measured over a period of many years (*typically over 30-years*). Climate can be thought of as the average patterns of weather in a given region.

From our daily lives, we are familiar with how forecasting influences our decisions: a glance at the weather forecast in the morning can determine our wardrobes for the day and even help us in deciding if and when we should leave our homes at all (and if we do, if we need to bring an umbrella). On vaster, and typically more consequential, scales, forecasting long-term environmental trends can (and should) determine socially significant planning. Climate predictions, for example, are relevant to the design of innovative national infrastructure and to disease abatement strategies [30, 31]. The act of forecasting can also shape the very future being predicted – we might call this *a hazard uncertainty principle* (that is, when a risky future is avoided by policy decisions made to avert that particular future). For this reason, environmental models are typically offered as a range of possible scenarios. A good environmental model presents a view of possible futures, ones that must be internally consistent with our knowledge about the dynamic behavior of the system. In forecasting more than one *possible* future, these scenarios provide a suite of alternative futures where our collective exposure to future loss or damage can be determined and assessed, and to which we must formulate a response.

Scenario development using quantitative models is only one way in which ecologists evaluate our potential risky future. Other ways of evaluating new and emerging risks include "horizon scanning" accompanied by expert evaluation of "what if" questions, along with the maintenance of risk registers. In a risk register, a catalog of potential calamities can be generated along with some provisional determination of management strategies in the face of an emerging risk [32].

A combination of scenario development, horizon scanning, and registration of risk invites us to ask: what sort of future *are we likely* to inhabit? What sort of future *do we want* to inhabit? What sorts of risks for life and well-being are associated with difference policy actions or inaction? How should these risks from environmental change best be measured? In order to address such questions as these, environmental students, more than ever, need the tools to think about risk: how is risk characterized, how are potential consequences assessed, how likely are the different outcomes? This framework whereby risk is characterized and evaluated forms the basis not just of environmental risk, but of *all* risk estimation.

1.4.1 Catastrophic Risk

In a now seldom read essay by Robert Louis Stevenson (1850–1894) entitled "Aes Triplex" (1878), the writer, more famous for his later stories *Treasure Island* (1883) and *The Strange Case of Dr Jekyll and Mr Hyde* (1886), wrote about the way in which we navigate the inevitable risks of being alive:

> We have all heard of cities in South America built upon the side of fiery mountains, and how, even in this tremendous neighbourhood, the inhabitants are not a jot more impressed by the solemnity of mortal conditions than if they were delving gardens in the greenest corner of England. ... It seems not credible that respectable married people, with umbrellas, should find appetite for a bit of supper within quite a long distance of a fiery mountain; ordinary life begins to smell of high-handed debauch when it is carried on so close to a catastrophe; and even cheese and salad, it seems, could hardly be relished in such circumstances without something like a defiance of the Creator. ("Aes Triplex", published in *Virginibus Puerisque*, 1881)

Robert Lewis Stephenson (1850–1894), best known for his gripping adventure stories was also an essayist: his best known essay is "Aes Triplex" (1878). In this essay Stephenson encourages us all to live our lives to the fullest and not to cower in the face of risk. He himself died young, turning to his wife as he was uncorking a bottle of wine saying, "What's that...does my face look strange?" After this he collapsed and was dead at age 44.

The question that this great essay poses for us is this: can we on the one hand objectively evaluate our risk and yet live with the sort of cheerful optimism that characterized that writer's demeanor? *Source:* Stevenson / Wikimedia Commons / Public Domain.

Stevenson – perhaps not coincidentally still a young man when he wrote this – cautions all of us against tepid living. He warns us as follows: "So soon as prudence has begun to grow up in the brain, like a dismal fungus," he writes, "it finds its first expression in a paralysis of generous acts." Stevenson's is therefore a call to live our lives without paying too much regard to the risk we run: "Who would find heart enough to begin to live," he concluded, "if he dallied with the consideration of death?" And yet might we not want, at the very least, to know something of the volcano, at the base of which we live, and project as best we can the environmental circumstances of the future in which we undertake our incautious adventures.

Environmental risk assessment is mainly concerned with risks at the higher end on the scale of implications: events that will have significant costs for life as well as for economic well-being. At the highest end of the scale of possible risky futures are rare events that are deemed truly catastrophic. In this context, war, famine, pandemics, and meteorite strikes (such as the one that caused the extinction of dinosaurs) might come readily to mind. However, more than ever before, risks associated with future upheavals of the environment are often categorized as catastrophic risks alongside the more frequently assessed global risks. It can be comforting to suppose that the future will be simply a linear extrapolation from present patterns – things will change, yes, and impacts may be exacerbated by inaction, but change should be steady and predictable. In recent years, scientists studying climate and other planetary processes have investigated concerns that future states for earth systems will be catastrophic *nonlinear* departures from recent trends. Such outcomes may manifest as "surprises" in the system. The magnitude of surprising outcomes can, by their very nature, be difficult to predict [33]. Be that as it may, surprises may include some highly consequential, low-probability, negative effects in climatic and ecological systems. For example, one fairly well-characterized climate "surprise" would be the shutdown of the Atlantic meridional overturning circulation (AMOC) – that is, the pattern of surface and deep currents in the Atlantic Ocean responsible for maintaining most of northern Europe in a relatively ice-free state [34]. There is some recent evidence that these currents *are* indeed weakening – if they were shut off it could result in catastrophic freezing of parts of the globe. After extensive evaluation, the ranking of this risk, on the one hand, is low at present, though, on the other hand, important thresholds in the activity of these currents may be closer than expected [35].

> The definition of **surprise** from a dictionary viewpoint is, well, not surprising: it refers to an unexpected occurrence; something that may astonish us. From an ecological point of view it means a radical and often not wholly prediction change in the property and behaviors of a system. Environmental scientists not only have to expect surprises in systems but need to offer advice on how to manage the world so as to avoid them.

Over the past decade, several high profile studies of potentially catastrophic surprises in climatic and ecological systems have been published (for examples of these see [36, 37]). These studies are often concerned with "tipping points" in the behavior of climatic and ecological systems. Thresholds exist is a wide variety of ecological systems: lakes, oceans, forests as well as in human societies, where a modest alteration may result in a potentially large change in system behavior though these thresholds are inadequately incorporated into risk models [38, 39].

> The concept of ***tipping point*** has been popularized in recent years. It refers to a state of a system where the behavior of that system whole becomes vulnerable to small changes that may cause radical and irreversible change. Tipping points are sometimes referred to as a "***critical transitions***" in the literature on systems theory.

1.5   The *Necessity* of New Ways of Thinking about Environmental Education

More than any other factor, it was a suite of papers on nonlinear surprises in earth systems published since 2010 that served, in their own way, as a tipping point in the way in which I taught, discussed, and wrote about environmental matters. My determination to incorporate risk frameworks into thinking about environmental change prompted the "conceptual turn" in my teaching and writing. Students need to know not only how to accumulate facts – as discussed above – but they *also* need robust ways of conceptualizing environmental risk, and understanding the real uncertainties that we face in the light of our evolving understanding of the (oftentimes nonlinear) behavior of environmental systems. Given the growing profile of environmental risk assessment within both academic and policy communities, and considering the potential catastrophic outcomes of some environmental change, we will therefore in a future chapter discuss a set of risk frameworks – called "catastrophic risk models" – that permit a comparison of environmental risk with other globally significant events: super-volcanoes, war, and pandemics, to name a few.

This emphasis on possible calamities – now commonplace in many environmental courses and books – can be anxiety inducing (something we shall also address in a future chapter) but it prepares students to confront and evaluate perspectives they will encounter in learning about this discipline, and, indeed, prepares them for some eventualities they may confront during their lifetime as citizens in the Anthropocene. To understand the world we inhabit and to act to fashion a safe, equitable, and flourishing future for all humanity demands an education commensurate with the scale of our task.

1.6   A Note of Alternative Conceptual Framing for Environmental Disciplines

The framing provided in this book is derived primarily from a Western scientific tradition (supplemented by the peculiarities of my own training as an Irish-born naturalist); the reader should be attentive to the emergence in recent years of a desire among many in the environmental community to anneal insights from Western scientific perspectives with indigenous scientific traditions. There are many indigenous scientific and knowledge systems, of course, but, in general, these traditions tend to be holistic rather than atomistic and reductive. Attempts to open a dialogue between world systems such as the Western scientific one and varieties of indigenous systems (sometimes referred to as Traditional Ecological Knowledge [TEK]) are appealing to many practitioners of ecological management [24, 40, 41]. However, such engagement is not without its difficulties. At least some indigenous scientists express concerns that the degree to which environmental science is at the service of economically extractive economic models (for medicine and even for techniques for the management of landscapes) is incompatible with indigenous worldviews, and can violate the intellectual property rights of local communities [42, 43].

> **Traditional ecological knowledge** (TEK) is the total accumulated knowledge concerning the relationship between traditional and indigenous cultures and their environments, as well as their assorted technical expertise regarding resources and the management of the land. Oftentimes, oral testimony, stories, and written records act as repositories of this knowledge [24, 44].

References

1 Leopold, A. (1989). *A Sand County Almanac, and Sketches Here and There*. USA: Oxford University Press.

2 Pörtner, H.-O. et al. (ed.) (2022). *Climate Change 2022: Impacts, Adaptation, and Vulnerability. Contribution of Working Group II to the Sixth Assessment Report of the Intergovernmental Panel on Climate Change*. Cambridge University Press.

3 Intergovernmental Science-Policy Platform on Biodiversity and Ecosystem Services (IPBES) (2019). Media release: nature's dangerous decline 'unprecedented'; species extinction rates 'accelerating' (May 5).

4 Brondizio, E.S. et al. (ed.) (2019). *Global assessment report on biodiversity and ecosystem services of the Intergovernmental Science-Policy Platform on Biodiversity and Ecosystem Services (IPBES)*. Bonn, Germany: IPBES Secretariat.

5 Redman, C.L. (1999). Human Impact on Ancient Environments. University of Arizona Press.

6 Jowett, B. (ed.) (2011). *Dialogues of Plato: Volume 2: Translated Into English, with Analyses and Introduction*. Cambridge University Press.

7 Goldin, O. (1997). The ecology of the *Critias* and platonic metaphysics. In: *The Greeks and the Environment* (ed. L. Westra and T. More Robinson), 73–82. Rowman & Littlefield.

8 Carone, G.R. (1998). Plato and the environment. *Environmental Ethics* 20 (2): 115–133.

9 Marsh, G.P. (1874). *The Earth as Modified by Human Action*. Ayer Company Publishers.

10 Steffen, W. et al. (2015). The trajectory of the Anthropocene: the great acceleration. *The Anthropocene Review* 2 (1): 81–98.

11 Bamzai, A.S. (2020). The NSF's role in climate science. In: *Oxford Research Encyclopedia of Climate Science*.

12 Shi, G. et al. (2021). Rapid warming has resulted in more wildfires in northeastern Australia. *Science of the Total Environment* 771: 144888.

13 Mann, M.E. and Gleick, P.H. (2015). Climate change and California drought in the 21st century. *Proceedings of the National Academy of Sciences of the United States of America* 112 (13): 3858–3859.

14 Sobel, A.H. et al. (2016). Human influence on tropical cyclone intensity. *Science* 353 (6296): 242–246.

15 Emanuel, K. (2005). Increasing destructiveness of tropical cyclones over the past 30 years. *Nature* 436 (7051): 686–688.

16 Donovan, M.K. et al. (2021). Local conditions magnify coral loss after marine heatwaves. *Science* 372 (6545): 977–980.

17 Doney, S.C. et al. (2012). Climate change impacts on marine ecosystems. In: *Annual Review of Marine Science*, vol. 4 (ed. C.A. Carlson and S.J. Giovannoni), 11–37.

18 Edwards, T.L. et al. (2021). Projected land ice contributions to twenty-first-century sea level rise. *Nature* 593 (7857): 74–82.

19 Boulding, K.E. (1956). General systems theory—the skeleton of science. *Management Science* 2 (3): 197–208.

20 Underwood, A., Chapman, M., and Connell, S. (2000). Observations in ecology: you can't make progress on processes without understanding the patterns. *Journal of Experimental Marine Biology and Ecology* 250 (1–2): 97–115.

21 Isopp, B. (2015). Scientists who become activists: are they crossing a line? *Journal of Science Communication* 14 (2): C03.

22 De Greene, K.B. (2012). *A Systems-Based Approach to Policymaking*. Springer Science & Business Media.

23 Von Bertalanffy, L. (1969). *General System Theory: Foundations, Development, Applications*. George Braziller.

24 Berkes, F. (1993). Traditional ecological knowledge in perspective. *Traditional Ecological Knowledge: Concepts and Cases* 1: 1–9.

25 Holmes, J. and Clark, R. (2008). Enhancing the use of science in environmental policy-making and regulation. *Environmental Science & Policy* 11 (8): 702–711.

26 Kindon, S., Pain, R., and Kesby, M. (2007). *Participatory Action Research Approaches and Methods: Connecting People, Participation and Place*, vol. 22. Routledge.

27 McKeon, R. (2009). Metaphysics. In: *The Basic Works of Aristotle*. Modern Library.

28 Lovelock, J.E. (1979). *Gaia: A New Look at Life on Earth*. Oxford Paperbacks.

29 Bush, A.B.G. et al. (2020). Issues in climate analysis and modeling for understanding mountain erosion dynamics. In: *Reference Module in Earth Systems and Environmental Sciences*. Elsevier.

30 Hall, J.W. et al. (2016). Strategic analysis of the future of national infrastructure. *Proceedings of the Institution of Civil Engineers – Civil Engineering* 170 (1): 39–47.

31 Escobar, L.E. et al. (2016). Declining prevalence of disease vectors under climate change. *Scientific Reports* 6 (1): 39150.

32 Sutherland, W.J. et al. (2021). A 2021 horizon scan of emerging global biological conservation issues. *Trends in Ecology & Evolution* 36 (1): 87–97.

33 Doak, D.F. et al. (2008). Understanding and predicting ecological dynamics: are major surprises inevitable? *Ecology* 89 (4): 952–961.

34 Rahmstorf, S. et al. (2015). Exceptional twentieth-century slowdown in Atlantic Ocean overturning circulation. *Nature Climate Change* 5 (5): 475–480.

35 Boers, N. (2021). Observation-based early-warning signals for a collapse of the Atlantic Meridional Overturning Circulation. *Nature Climate Change* 11 (8): 680–688.

36 Kerr, R.A. (2008). Climate tipping points come in from the cold. *Science* 319 (5860): 153–153.

37 Lenton, T.M. (2012). Arctic climate tipping points. *Ambio* 41 (1): 10–22.

38 Scheffer, M. et al. (2001). Catastrophic shifts in ecosystems. *Nature* 413 (6856): 591–596.

39 Scheffer, M. (2020). *Critical Transitions in Nature and Society*. Princeton University Press.

40 Robinson, J.M. et al. (2021). Traditional ecological knowledge in restoration ecology: a call to listen deeply, to engage with, and respect indigenous voices. *Restoration Ecology* 29 (4): e13381.

41 Kimmerer, R. (2013). *Braiding Sweetgrass: Indigenous Wisdom, Scientific Knowledge and the Teachings of Plants*. Penguin Books.

42 Isaac, G. et al. (2018). Native American perspectives on health and traditional ecological knowledge. *Environmental Health Perspectives* 126 (12): 125002.

43 Gupta, R., Gabrielsen, B., and Ferguson, S.M. (2005). Nature's medicines: traditional knowledge and intellectual property management. Case studies from the National Institutes of Health (NIH), USA. *Current Drug Discovery Technologies* 2 (4): 203–219.

44 Kimmerer, R.W. (2002). Weaving traditional ecological knowledge into biological education: a call to action. *BioScience* 52 (5): 432.

2

A Conceptual Approach to Environmental Science

Thoughts without content are empty, intuitions without concepts are blind

Immanuel Kant

Two approaches to doing science can be distinguished: one involves the relentless accumulation of particular facts before declaring a universal law (*induction*); the other uses the "searchlight" of theory to direct our exploration of the world to test these theories (this is sometimes called – grandiosely – the *hypothetico-deductive model*).

The analogy of the searchlight can be applied to the *role of concepts* in the environmental sciences. The conceptual approach invites us to use a shared mental representation of the world – for example the concepts of *environment*, or of a *species*? – to frame our encounters with the real world of nature. Insights can then be shared with others to develop hypotheses and undertake research.

Try this: identify a number of concepts that you regularly use – cat, tool, cake… – and see (i) is your concept shareable (do you use the concept in the same way as others)?, (ii) can this be used as searchlight to spot similarities and differences between objects in the real world?

Source: Theodor Severin Kittelsen / Wikiart / Public Domain.

A Primer on Human Impacts on the Environment: The Conceptual Approach, First Edition. Liam Heneghan.
© 2023 John Wiley & Sons Ltd. Published 2023 by John Wiley & Sons Ltd.

The world is awash with facts; these facts are categorized by our *concepts*. Concepts – that is, the mental representation of the properties of things – can serve to frame our encounters with the world; they allow us to assemble facts into a coherent picture. How concepts, when they are assembled into sophisticated frameworks, enable us to both grasp the immensity and complexity of global processes, while also simplifying this complexity sufficiently to allow us to understand the world, is the subject of this chapter. But, first, let us describe in quite simple terms what a concept is and why concepts form the bedrock of thought.

Concepts are important not only when we are undertaking scientific work, but also in our everyday dealings with things. We rely upon relatively stable concepts of particular entities – the concept of a *cat*, say, or of a *flowerpot* – in order to unify those experiences that endure for more than a moment. Let us imagine that I have been writing these sentences in a room where the cat that had been crouching behind the flowerpot on a shelf above the door leaps down upon the sofa right behind me. Now, let us suppose that you have been reading these lines in a room where *your* cat leaps from the sofa to the shelf above your door and upsets the flowerpot. In each case, the cat remains very much *a cat* through all of its escapades; that is, the *conceptual cat* endures as these *particular cats* bound about their respective rooms. Furthermore, our shared concept of *cat* (and sofa, and flowerpot) permits you to understand me as I share this anecdote. As you brought these cat vignettes to mind, did you, reader, imagine an agile feline gallivanting about a furnished room? If so, this is the communicative power of concepts: without concepts we cannot share our experiences.

Concepts, as we have already implied, can be combined to produce a complex picture of the world. Several constituent parts can be assembled into a single conceptual *system*. The cat, the sofa, the shelf, that flowerpot, the door, the room, and my observing it all, and even you imagining this scenario, are part of the one complex framework. We use concepts, both simple ones and complex ones, all the time, without being fully aware of it. On the grander scales that we consider in this book, the earth in its entirety is part of a complex system. Specific parts of the systsem come into sharp focus as we view the world through the lens of a particular conceptual frameworks. For example, when employing the concepts of "environment," "biosphere," or "climatic systems" different aspects of our world are highlighted.

In this chapter, I make a case for the utility of conceptual models as a way of organizing our factual accounts about the changing environment. Abstract frameworks act as a searchlight in directing our observations of and framing solutions to the problems associated with human impact on globally-scaled phenomena. We will proceed as follows: initially we investigate how concepts are employed in everyday practice. To illustrate this, we will first consider a very simple example that gives us the gist of how we use concepts to collate facts, and then we will consider a more complex example that shows how different conceptual frameworks have, throughout history, organized the human understanding of *the world*. That is, we ask how has the concept of *world* framed the way in which different cultures relate to their surroundings? Finally, we will distinguish a *conceptual model* from a *mental model*, showing the fundamental role of each in both learning and in guiding research. Equipped with these fundamentals we turn in subsequent chapters to some of the more important conceptual tools for environmental thinking.

2.1 Facts and Concepts: A Simple Environmental Example

Phenology is the study of the timing of seasonally recurring natural phenomena. This discipline engages in observations on the timing of events associated with all sorts of organisms. Particular attention is paid, however, to the timing of both leafing out and flowering of plants, and to the dates of the return of migratory animals. The accumulation of phenological data can be helpful in monitoring climate change. Such efforts can be the basis of excellent citizen-science projects (e.g. the Project Budburst, managed by the Chicago Botanic Garden, collates data on plant seasonal events and on insect pollination visitors to plants. See https://budburst.org)

Today's temperature in Chicago, Illinois, is a mild 52°F (11°C); the days are getting longer (sunset is expected at 7:06 p.m.). Outside, I observe that the snowdrops are already flowering, the buds on the Magnolia shrubs are getting stout, and though the ground remains cold, the daffodils bulbs in the garden have sprouted; the earthworms are casting where the soil is soft and muddy. Along the shores of Lake Michigan, the Red-winged blackbirds are feistily preparing their territories for the mating season. Putting these observations together I can declare that, at last, spring has returned to the US Midwest. The role that *facts*, such as the ones I have just observed, perform are various. Facts are components of true sentences: when I declare that "the temperature is 52°F", this communicates a verifiable truth about the temperature. We can say, then, that facts, declare "actual states of affairs," and that they are "true propositions." Armed with the facts we can investigate causal relationship between facts, asking in this case, for example if the cue for the flowering of the snowdrops is the extended hours of sunshine or an elevation in temperature due to encroaching spring? In addition to describing "actual states of affairs," facts can thereby serve to unravel causal relationships. Because of this, facts are foundational for scientific research. All in all, the factual conditions that I have describe above are representative of this time of year (today's date is 23rd of March) – it's a typical early spring day for the region in which I live. Now, you might ask, why have I assembled these *particular* scraps of information? On the one hand, they are just a random set of observations helpful in illustrating what is meant by the word "fact." But I have more in mind than this. I am interested in this constellation of facts in particular because I am seeking to confirm scientific reports that the date of the last spring freeze in the US Midwest occurs at least one week earlier than it had in the early years of the twentieth century [1]. A recent study of the timing of egg laying by 72 bird species from the Midwest, confirms this impression of the early arrival of spring, by revealing that dates of egg laying has advanced by about 10 days over the past century or so [2]. Winter in Chicago has become just a little shorter. The collection of facts that I noted above can now be seen as not random at all: they are phenologically relevant observations useful in evaluating climate change in the Midwest. These phenological facts also illustrate the relationship between theory and data: that is, the importance of these phenological indicators of spring becomes clear when assembled in the light of our concerns about climate change. To employ an analogy attributed to Austrian philosopher of science Karl Popper (1902–1994), we have used expectations about climate change as a "searchlight" to guide us to facts concerning changes to timing of the seasons. To conclude this example concerning phenology, I must add that in order to be scientifically meaningful, such facts would need to be systematically evaluated on a larger scale than I have sketched here – that is, climate is assessed over a long period of time, typically 30 years, whereas weather refers to day-to-day variation in meteorological conditions.

Weather Extremes and Climate Change

Since weather is the day-to-day variation in meteorological conditions whereas climate represents longer term meteorological patterns, it is not always helpful to use singular anomalies in the weather to speculate on climate change. That being said, over the past decade scientists have developed a range of tools (called attribution tools) that assess the probability of extreme weather events under different scenarios of climate change. These tools allow for a statistical evaluation of weather extremes, an approach that is especially helpful in informing disaster planning and assessment.

Philosophy of science: *some handy terminology*

Hypothesis: a conjecture developed to account for the observed facts about a natural phenomenon. Often these are a starting-point for a scientific study.

Induction: the processes of inferring a general law based upon observations of a number of particular instances.

Hypothetico-deductivism: the use of hypotheses to formulate expectations about the behavior of a system. The expectations should be logically consistent with (or deduced from) the hypothesis. If the data confirm the deduction, the hypothesis can be tentatively accepted.

Falsification: according to the philosopher Karl Popper, the only statements that may be considered truly scientific are those that are open to falsification – that is, any conjecture that cannot in theory be refuted is unscientific. One implication of this premise is that scientific observations must always be sought that attempt to disprove hypotheses, our current thinking is thus always provisional and may be shown in the future to be false.

Theory: a conceptual device for systematically characterizing the behavior of systems. Often, a well-developed scientific theory provides laws which adequately describe the features of a complex system (e.g. atomic theory, evolutionary theory).

The powerful analogy of the "searchlight" relates to one of Popper's distinctive contributions to an understanding of the logic of scientific discovery. Popper's claim is that, contrary to an assumption people often hold, scientists do not simply accumulate disparate facts and from these observations then, and only then, construct a theory of how the world works (often called the inductive method). The inductive approach seems to imply that the mind is a *tabula rasa*, a blank slate upon which experiences leave their impression without any contribution from the concepts. Contradicting this, Popper sees the process as reversed – concepts guide the accumulation of knowledge. The sciences use well-constructed conjectures (or hypotheses) about the workings of the natural world to determine which facts are relevant for testing these theories. Thus, theories and hypotheses – and the foundations upon which these mental constructs are based – are used as searchlights to direct an observer to facts that can then, in turn, test theory. To conclude our brief encounter with Popper's thought, we should add that the facts that we have assembled in light of theory can then be used to potentially "falsify" our theories – thus the falsification of a theory occurs (in principle, though not always in practice) when observations made in the light of theory are consistently contrary to those expectations.

2.2 The "World" as Concept

We turn next to a more complex conceptual example – this time it is an example that is on the scale of that entity central to this book's enquiry: that is, the scale of planet in its entirety. The terms "world" and "planet" are often used as near synonyms in the powerful, though circumscribed, epistemology of Western science. The notion of "world" prevalent in the environmental sciences is that our world is a 4.5 billion year old rocky body that revolves in an approximately elliptical orbit around the sun and that supports living things including ourselves; that is, the world of environmental science is this physical Earth. This is a relatively recent conception of the world, details of which have developed (for the most part, at least) since the dawn of the Scientific Revolution in the sixteenth century – and it is a conception of the world that has become increasingly complex and complete over the centuries [3–5].

Beyond this prevalent use in the contemporary natural sciences – and the world as Earth is the version that we will primarily employ in this book – the concept of the "world" is, historically, an almost endlessly rich one. The role of the concept of "world" in the ancient cosmology of India, Mesopotamia, Greece and the Hellenistic world, Egypt, China, Mesoamerican, and Andean cultures, to give several examples, has attracted significant scholarly attention [6]. Reflecting upon the conceptual variations of "world" expressed in philosophical and theological work and in the cosmologies of diverse cultures is a useful exercise. That is because it reveals the way in which intricate conceptions of "world" have always served to describe central aspects of the cosmos in different places and times, and determines action for members of that particular community. In essence, we have always encountered facts (that is, "true propositions") with the help of culturally appropriate worldviews. The facts we notice become known in the light of a worldview.

To choose just one from among the many examples, the ancient Mesopotamian conception of "world" has been reconstructed from those mythologies, temple hymns, and inscriptions associated with ziggurats that have come down to us from the earliest eras (from approximately 3100 to 539 BCE [the fall of Babylon]). The concept of "world" expressed in these sources is the *entirety of creation* and can be divided into two principle realms: heaven and earth (with an accompanying idea of an underworld) [7, 8]. The Mesopotamian world was fashioned by Marduk – the chief god of Babylon whose cult was especially prevalent in the eighteenth century BCE – who split the carcass of the sea-mother Tiamat to create, with one part, the region of heaven, and with other parts the many aspects of the natural world. Tiamat's eyes, for example, are regarded as the source of the Euphrates and the Tigris [9]. In Mesopotamian cosmology humans are accorded a special place – humans occupy one of the three layers of Earth. Their special role notwithstanding, humans ultimately are beings living their lives in the service of the gods [8].

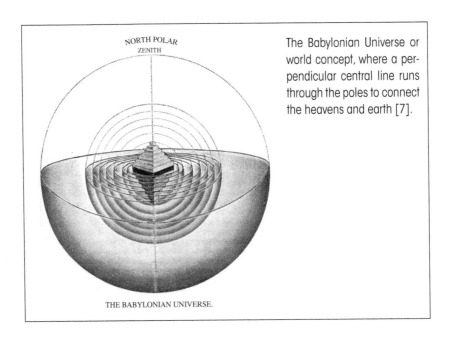

The Babylonian Universe or world concept, where a perpendicular central line runs through the poles to connect the heavens and earth [7].

THE BABYLONIAN UNIVERSE.

It is difficult for us mortals – Mesopotamian or contemporary – to grasp the world as a whole. Thus, every conception of the world should be regarded as a partial one. As we saw with our simpler example – concerning the phenological evidence of spring – certain aspects of the world are highlighted in the glare of a particular conceptual searchlight, while others remain occluded. It is by means of the necessary simplifications associated with conceptual development that each individual worldview draws its strength. Because of this, partial representations of the world illuminate some aspects of the world while also posing specific challenges to interpretation. Models of the "world" are elaborated to serve a particular set of purposes (even when, as is the case with Western science, a worldview aspires to objectivity [10, 11]). The Mesopotamian schemata – of which we have considered but a small fragment – provides an account of the creation of the cosmos, while also codifying the role that humans play in it. According to Edwin James (1888–1972) a scholar of comparative religions, in distinction from natural scientific frameworks, mythological accounts of the world serve primarily to "validate and justify, conserve and safeguard the fundamental realities and values, customs and beliefs on which depend the stability and continuance of a given way of life" [12]. A modern environmental understanding of our world – that is, "world" understood as this planet Earth and the living beings it supports – also provides a material account of the place of humans in the grand scheme of things (though such an account may also have spiritual implications [13]). This concept of the "world" can help us understand how the planet is maintained as suitable habitat for humans, as well as pointing to those transformation of planet that might be adjudged harmful. The conceptual tools of environmental sciences may also seek to assist in repairing the damage.

In this volume, the concept of the "world" that we investigate consists of two multifaceted and interacting components. On the one hand we have the astronomically significant nonluminous satellite upon which humans and other species live. That is this physical planet Earth. To understand even the most basic physical phenomena of *this* realm necessitates a familiarity with, for example, geology, oceanography, climatology, and soil science. Each of these disciplines, in turn, has a vast compendium of conceptual tools at its disposal. Such models direct the research efforts of scientists engaged in these subdisciplines and assist in identifying gaps in our understanding. Thus, the conceptual apparatus is made of complex interlocking parts, each of which is the business of a specialized science. The other major component in the scientific conception of the "world" concerns the 200,000-year-old primate, that is, humans, who are now almost ubiquitous on the planet. To fully appreciate the role of humans and other living beings on the planet requires an understanding of biology – with all its subdisciplines – and the ecological sciences, which in their turn have a complex intellectual history [6]. Furthermore, to fully appreciate humans and their capacity to shape this planet requires the resources of the social, cognitive, and behavioral sciences. The task of connecting primate and planet (that is, bringing the two components of the scientific conception of the world together) is a hugely interdisciplinary one, with all of the prospects and challenges inherent in interdisciplinary studies [14]. Yet, despite the multidisciplinary undertaking of the environmental sciences, its "world" still remains a partial one in the sense that it often confines its attention primarily to humans and the most important conditions that sustain us.

To simplify an otherwise daunting undertaking, our approach here in this book will be to examine those conceptual models that have proved most salient for comprehending the transformation of the world since humans took up their tenancy on the planet. These transformations throughout the course of history were, and sometimes are still, undertaken with great deliberation; at other times humans have quite unintentionally changed the planet as we transform it to suit our own needs [15]. While we have shaped the planet, the planet has, in turn, shaped us. Other books discuss this latter transformation in considerable detail but on the question of the influence of the planet on the primate we will remain largely silent except where it is directly relevant to understanding global transformations [8].

2.3   From Concepts to Models: Important Terminology

Concepts, as we have now seen, are the mental representation of the properties of things that can be assembled in complex ways to inform our understanding of the world (the Mesopotamian world view, for example, or that of contemporary science). In this way, concepts form the bedrock of our higher thought processes. In addition to facilitating our efforts to make sense of the world, conceptual schemes act as a searchlight in helping identify gaps in our knowledge.

For clarity, I will name our complete cognitive grasping of how individual concepts are linked to provide a representation of an aspect of the world our "mental model" of that phenomenon. Mental models accomplish cognitive tasks and they play an especially important role in the application of knowledge to solving problems [10, 16].

An understanding of mental models is important in teaching practice. An instructor's task is often to supplant students' simplistic (and often incorrect) models with more powerful and empirically based ones [17, 18]. For example, a young child will often have a very rudimentary understanding of their own body. The child might, for instance, extrapolate from their intuitive understanding of their ears to imagine that a simple internal tube connects the visible parts of both ears. After all, each ear canal seems to be directed inwardly – why should they not connect? The teacher (or guardian) must introduce the complicating fact that the brain and other cranial tissues interpose between the ears, and in this way the simple model is dislodged and a more sophisticated model supplants it. As the child's understanding of their own body becomes increasingly complex, they may, with the assistance of anatomical drawings, and so forth, eventually gain a highly complex understanding of how their auditory systems works. Misconceptions about the human body can be persistent, and supplanting faulty concepts with ones that are more appropriate for medical practice can be challenging even for medical educators [19, 20].

Understanding the mental models that people have about the biophysical workings of the planet have implications for their perception of environmental problems like climate change, and conservation issues [21, 22]. It is clear, however, that misconceptions about basic environmental concepts, and about specific problems such as climate change, represent a challenge for environmental education [22, 23]. In ideal circumstances, the role of the instructor is to allow students to understand and correct their intuitive models, illustrate how incorrect models can make it difficult for the student to make sense of environmental phenomena, and initiate the processes of correcting these models. Though the process of model correction is easy enough to grasp in theory, in reality, some of our mental models – in diverse fields of enquiry: from astronomy and zoology – can be exceedingly tenacious [24].

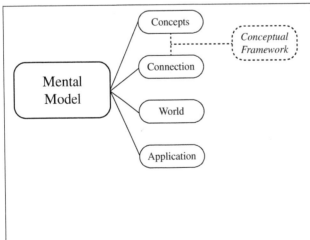

The relationship between the "conceptual frameworks" and "mental models" [2]. We will call a depiction of the conceptual frameworks in graphical form we are calling a conceptual model. Our mental models integrate concepts with the processes that connect these concepts, and help us place these alongside an understanding of the world. Furthermore, mental models are supplemented with an conjecture about the way in which this information can be applied in a variety of real-world situations. Illustration by Sarah Heneghan.

It can often be helpful for the development of research projects to draw a diagrammatic representation of models: indeed it is often a requirement of research proposals to include a *conceptual model* (i.e. a graphical representation of the researchers understanding of the phenomena they intend to work on [17, 25].) In this view of things conceptual models illustrate the subcomponents of mental models dealing with concepts and the processes that connect them.

2.4 Components of Successful Conceptual Models

Conceptual models that encapsulate our mental representations of particular concepts and the connections between them can serve to help us in explaining the phenomena of nature. They are important as tools for both framing research and in teaching. In this section, we set out some rules for successfully depicting conceptual models. (Though terminology varies among different authors (and "framework" and "model" are often used interchangeably), here I refer to the graphical depictions of conceptual frameworks as conceptual models).

A successful conceptual model will have the following features.

1) *Simplification*

   Statistician George Box once wrote, "All models are wrong, but some are useful" [26]. This is true, in part, because a model, by necessity, is a simplification of a system designed to make a problem, or phenomenon, tractable. After all, as mathematician Eric Temple Bell, quipped "the map is not the thing mapped" [27]. That is, in order to be useful, the model must only embed a level of detail sufficient to highlight those features of the world that are worth investigating. Encapsulating this idea, John Harte, an environmental scientist and policy scholar, conveys the idea of necessary simplicity wittily in the title of his book on problem solving in environmental science: *Consider a Spherical Cow: A Course in Environmental Problem Solving* (1985) [28]. Cows, of course, possess complex geometrical shapes, but for computational ease imagining the intricate bovine shape to that of a sphere can be a helpful simplification (at least for some purposes).

An image of a spherical cow used as the program cover of the 1996 annual meeting of the American Astronomical Association. *Source:* NASA / Wikimedia Commons / Public Domain. http://www.ikallick.com, last accessed July 14, 2022.

2) *Knowledge Gaps*:

   Models can be helpful in both illustrating hypotheses but also in identifying knowledge gaps. Oftentimes a researcher might have confidently identified those key processes that link entities or concepts in a model without knowing the precise mechanisms underpinning the interactions. Inspection of the model can facilitate research on feedbacks and interaction strengths in the system under investigation, and can sharpen the focus of future research.

3) *Refinement and Modularity*

A good conceptual model should be amenable to refinement. Refinements can emerge when investigators reconcile their models in collaboration with others who work on shared problems [25]. When new information becomes known – in response to research inspired by the model, or by general advancements in the field – the model should be adaptable to these circumstances. New conceptual elements, new processes, refined conceptions of feedback, and better characterized interaction strengths can be incorporated into refined models. Additionally, the model can be applied to new problem sets which may necessitate a reconfiguration of the basic form of the model. Ultimately each specialized model should articulate, as a model, of progressively more complex models the provide a complete representation of our understanding of the world.

4) *Interpretive Frameworks*

Ultimately, conceptual frameworks present an interpretation, or a lens for understanding a given system. Unlike purely mathematical models which may generate quantitative predictions, conceptual models sometime have "softer" aspirations: that is, the development of a generalized understanding of system behavior. Though many conceptual frameworks are quantitatively informed, some are purely qualitative research tools. Undoubtedly, quantitative data can be generated when research is guided by outlining a conceptual framework, even when this model is developed qualitatively.

In order to employ a conceptual model in learning or for use in research, the model should draw the attention of the student or investigator to the relevant sources of information about domain being investigated. For this reason, model development is important in the writing of research proposals, and indeed can often be a requirement for grant applications, or in graduate dissertation proposals. The exercise of identify important gaps in knowledge, can make the investigator aware of sources of existing data, or can help in designing the most appropriate experiments or observations.

Viewed in this way, we can see that a conceptual framework can never completely encapsulatate a phenomenon. The investigator must always be open to revising these models, and to recognizing that a novel observation or theory may make their framework obsolete. Then the process of model development resumes.

2.5 Coda

Undoubtedly, the organization of concepts and their presentation in the form of conceptual models are integral parts of the environmental disciplines [25, 29]. Recent developments of environmental models have helped focus research and policy attention on damage to the health of ecosystems on the global scale, while also encapsulating the dependency of humans upon the goods and services of nature for their sustenance. Many of these conceptual models are considered to be grounded in a more general systems theory. We will turn to these models in the next section of the book. However, before considering these systems models, in the next chapter we will briefly discuss the role of definitions in environmental thinking. Formulating and refining definitions is often underappreciated: but the matter cannot be ignored since many of the key concepts in this discipline – environment, species, ecosystems, for example – are notoriously resistant to definition.

References

1 Hayhoe, K. et al. Climate change in the Midwest. In: *Confronting Climate Change in the US Midwest*. Union of concerned scientists http://ucsusa.org/assets/documents/global_warming/midwest-climate-impacts.pdf.

2 Bates, A.E. et al. (2014). Defining and observing stages of climate-mediated range shifts in marine systems. *Global Environmental Change-Human and Policy Dimensions* 26: 27–38.

3 Longair, M.S. (2004). *A Brief History of Cosmology*, vol. 2. Citeseer.

4 Rees, M. (2000). *New Perspectives in Astrophysical Cosmology*. Cambridge University Press.

5 Bowler, P.J. (1993). *The Norton History of the Environmental Sciences*. WW Norton & Company.

6 Motz, L. and Weaver, J.H. (1995). *The Ancient Cosmologies*, 13–19. Springer.

7 Lambert, W. (1975). *Cosmology of Sumer and Babylon*. In: *Ancient Cosmologies* (ed. C. Blacker and M. Loewe), 42–65. George Allen& Unwin.

8 Rochberg, F. (2020). Mesopotamian Cosmology. In: *A Companion to the Ancient Near East*, 2e (ed. D.C. Snell), 305–320. Wiley.

9 Geyer, J.B. (1987). Twisting Tiamat's tail: a mythological interpretation of Isaiah XIII 5 and 8. *Vetus Testamentum* 37 (2): 164.

10 Meyer, H. (1951). On the heuristic value of scientific models. *Philosophy of Science* 18 (2): 111–123.

11 Ziman, J. (1996). Is science losing its objectivity? *Nature* 382: 751.

12 James, E.O. (1957). The nature and function of myth. *Folklore* 68 (4): 474–482.

13 Berry, T. (2009). *The Sacred Universe: Earth, Spirituality, and Religion in the Twenty-First Century*. Columbia University Press.

14 Frodeman, R., Klein, J.T., and Pacheco, R.C.D.S. (2017). *The Oxford Handbook of Interdisciplinarity*. Oxford University Press.

15 Tattersall, I. (2012). *Masters of the Planet: The Search for our Human Origins*. St. Martin's Press.

16 Greca, I.M. and Moreira, M.A. (2000). Mental models, conceptual models, and modelling. *International Journal of Science Education* 22 (1): 1–11.

17 Pundak, D., Liberman, I., and Shacham, M. (2017). From conceptual frameworks to mental models for astronomy: students' perceptions. *Journal of Astronomy & Earth Sciences Education (JAESE)* 4 (2): 109–126.

18 Gruber, H.E. and Vonèche, J.J. (1977). *The Essential Piaget*. Routledge and Kegan Paul London.

19 Bordes, S.J. et al. (2021). Medical student misconceptions in cardiovascular physiology. *Advances in Physiology Education* 45 (2): 241–249.

20 Feltovich, P.J., Spiro, P.J., and Coulson, R.L. (1988). The Nature of Conceptual Understanding in Biomedicine: The Deep Structure of Complex Ideas and The Development of Misconceptions. *Center for the Study of Reading Tech. Rep. 440.*

21 Mäki, U. (2001). Models, Metaphors, Narrative, and Rhetoric: Philosophical Aspects. In: *International Encyclopedia of the Social and Behavioral Sciences* (ed. N.J. Smelser and B. Baltes), 15–9931. Elsevier.

22 Munson, B.H. (1994). Ecological misconceptions. *The Journal of Environmental Education* 25 (4): 30–34.

23 Gibson, D.J. (1996). Textbook misconceptions: the climax concept of succession. *The American Biology Teacher* 58 (3): 135–140.

24 Klein, G. and Baxter, H.C. (2006). Cognitive transformation theory: contrasting cognitive and behavioral learning. Paper presented at Interservice/Industry Training Systems and Education Conference, Orlando, FL (December 2006).

25 Heemskerk, M., Wilson, K., and Pavao-Zuckerman, M. (2003). Conceptual models as tools for communication across disciplines. *Conservation Ecology* 7 (3).

26 Box, G.E. (1976). Science and statistics. *Journal of the American Statistical Association* 71 (356): 791–799.

27 Korzybski, A. (1958). *Science and Sanity: An Introduction to Non-Aristotelian Systems and General Semantics*. Institute of GS.

28 Harte, J. (1988). *Consider a Spherical Cow: A Course in Environmental Problem Solving*. University Science Books.

29 Pavao-Zuckerman, M.A. (2000). The conceptual utility of models in human ecology. *Journal of Ecological Anthropology* 4 (1): 31–56.

3

A Short Chapter on the Definition of Definitions

A difficulty for environmental studies, in general, is that the discipline is concerned with conceptual entities that are notoriously resistant to definition. The concept of "nature" itself is so all encompassing that scholars treat the term by either adopting a historically nuanced approach [1] – specifying the multitudinous ways in which the word has been employed over time – or by abandoning the term altogether [2]. The term "environment," which seems more contained than "nature," is also (as we shall explore further in *Chapter 6*) itself vague in the sense that without some modifier (forest environment, chemical environment for example,) it is difficult to operationalize the concept [3]. How, one might wonder, can we measure or account for something that seems to simply mean "the surroundings" [4]? Though this problem of vague definition is felt acutely in the environmental disciplines, the difficulty is not unique to these disciplines. Most of those natural and social sciences dealing with complex and multifaceted entities struggle to achieve consensus definitions of their foundational concepts. For example, the concept of "life" in biology, "culture" in anthropology, "society" in many social sciences are all notoriously contested terms [5–7].

In what follows, we reflect somewhat philosophically on the properties of definitions. Arguably, debates over the nature of definitions have been central to the Western tradition of enquiry since the time of the Ancient Greeks. Although it may well be that identifying that one true *essence* of the word "definition" will continue to elude us, nonetheless, the sciences are founded upon the task of defining things. Definitions provide a foundation for research, but are also a product of scientific work, since the sciences often serve to identify previously unrecognized entities. Entities from black holes and quarks in physics to ecosystems and the biosphere in environmental science are relatively novel, and recognizing and defining them can then fundamentally alter the way in which we see the world.

A Primer on Human Impacts on the Environment: The Conceptual Approach, First Edition. Liam Heneghan.
© 2023 John Wiley & Sons Ltd. Published 2023 by John Wiley & Sons Ltd.

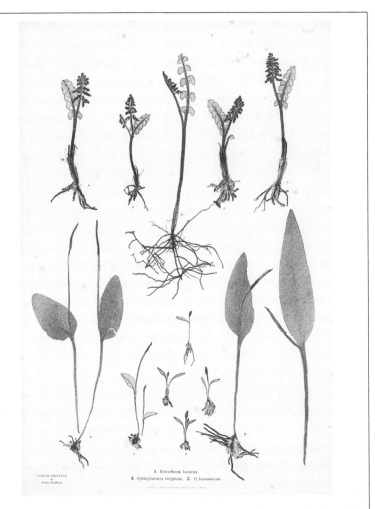

This herbarium sheet preserves parts (and wholes) of individuals from different species of fern. Botanists use observations on many individuals, with all their variety, in an attempt to define the essential properties of each species. The taxonomist, in this way, alters us to the limits of variation in each species, and announces where to place boundaries in flux of nature. Nature Print of Moonwort and Adder's Tongue Ferns, plate 51 from *The Ferns of Great Britain and Ireland*. 1855. Henry Bradbury. *Source:* The Art Institute of Chicago / CCO 1.0.

3.1 A Definition of Definition

We have seen that concepts can be simple (cat) or complex (world). Specifying the precise meaning of a concept, however, can be a difficult task. A description of the content of a concept is called its definition. Though its importance is rarely explicitly acknowledged, the task of providing definitions is a basic preparatory step in doing any science, and, to some extent, it will preoccupy us throughout this book. To define something is to set bounds or limits on that thing. Consistent with its boundary-setting objective, the etymology of "define" is from the Old French word "definer" meaning to end or to terminate [4]. When we define any entity we, in effect, determine where that particular thing ends and where another thing begins. In this way, a definition renders a thing (or class of things) distinguishable from everything else in the universe. For a definition to perform its discriminatory function it must, at the very least, provide a good description of the properties of a thing. In some philosophical accounts of definitions, the attributes of objects are considered to derive from an entity's *essence* (or *essential nature*) [8]. Scientific definitions in particular aspire to formally identify the essential properties of things (and often express these identities mathematically).

Recognizing the foundational importance of definitions may seem intuitive, yet determining the desirable attributes of a precise definition is a complicated matter. Though these issues can seem arcane, it is a particularly important one for environmental thought, since, as we will discuss in a later section, many of the disciplines most basic terms – for example, "nature" (as we have seen) and even "environment" – are highly contested terms, the definitions of which vary greatly.

Classic examples of the difficulty posed by definitions can be found in the early Socratic dialogues [9]. In the *Euthyphro* dialogue (circa, 399 BCE), for example, Socrates in his inimitable (though admittedly irritating) style seeks a definition of "piety." Having such a definition was a matter of life and death for Socrates since he was facing trial for impiety among other supposed crimes. "What is the pious, and what is piety?" Socrates asked Euthyphro (a young Athenian religious leader). In one of his numerous attempts at a reply, Euthyphro suggests to Socrates that an action is pious if it pleases the gods (this may conform, even today, with a notion that piety involves displaying reverence to God or the gods). Does that mean, Socrates then wondered, that piety is determined simply by the (perhaps whimsical) judgment of the gods, or is there something inherent, that is, some *essence* of piety, that determines a pious action, an essence that, in turn, results in divine satisfaction? In his interrogation of the younger, theologically inclined, Euthyphro, Socrates is clearly soliciting a particular form of definition. The definition Socrates seeks evidently cannot emerge merely by presenting a list of several instances of piety (that is, he rejects an ostensive form of definition, where examples of piety are enumerated). Ostensive definitions – which simply point out instances of the thing being defined – fail to get at the essence of a term. An essential definition of a thing is one against which all potential instances of piety may be adjudged.

It can be difficult to identify the essence of an entity; ultimately, Socrates gleans very little from his conversation with Euthyphro (and the dialogue ends in a philosophical *aporia* – that is, a state of perplexity). The notion that all things have essences, or primary qualities – to use a term often used in philosophical discussions of definitions – as well as possessing nonessential or "accidental" properties, has remained vexatious throughout the history of Western philosophy [10, 11]. To illustrate this distinction: ponder the definition of the cat (the one we met in the previous chapter). We might be unsatisfied if our interlocutors described representative colors of cat fur; we are typically more interested to know the essence of the feline nature. In ordinary conversation we may be satisfied with a "lexical" definition of a cat [12]. Lexical definitions are those typically found in a dictionary: for example, we can define a cat as "a well-known quadruped. . . often kept as a pet, and generally known to be skilled at catching mice." This is good enough for ordinary conversation but is perhaps not good enough for doing science, where we might resort to genetic studies of domesticated cats. It is a matter of debate whether in gleaning genetic information about cats we found some purely essential properties (in a philosophical sense) but surely the genome provides a basis for scientific work.

For our scientific purposes, we must certainly seek definitions that distinguish the borders or limits of that item being defined: we want to be able to delineate. Along with Socrates, we will want ensure that we are identifying primary rather than superficial qualities of the entities and concepts that we discuss. It is worth reflecting upon the fact that Socrates retired from his dialogue with Euthyphro without gaining a satisfactory definition of piety. He was subsequently put to death for "impiety" and "corrupting the young," that is, he died without anyone offering a satisfactory definition of his crime. We will not, I hope, be facing such a grim fate as this but nonetheless we should not always expect to discover overly strict forms of definitions. At the very least, we need to be appropriately humble when we reach a moment of puzzlement, or aporia.

Socrates, in his early dialogues, made the proposing of definitions a central activity in the task of knowledge creation. His discovery was that few of his interlocutors could provide compelling definitions of commonly used terms. A consequence of these Socratic insights is that we should pause to reflect upon the meaning of those concepts that are central to our moral lives. *Source:* Derek Key / Flickr / CC BY 2.0.

3.2 Role of Definitions in Scientific Explanations

The definition of a definition is ultimately simple enough: it is "a statement, declaration, or proposal establishing the meaning of an expression" [10]. Agreeing that we have found a meaning of an expression or concept under consideration sufficient for providing a definition is, however, not a simple matter. This is because in addition to what we have already considered, what serves as a good definition depends upon context. In a conversation with a friend, a definition can be contextual. Your friend may simply understand an expression from the context of your exchange. In contrast, scientific studies usually require an explicit definition that directly fixes the meaning of a term.

Explicit definitions are a *prelude* to a scientific activity. However, definitions may also be *end products* of science as much as they are prerequisites for undertaking it [13]. That is, a scientific investigation may reveal that an entity formerly thought to be singular should actually be considered multiple. For example, recent genetic studies reveal that in several instances populations formerly classified as belonging to the same species are, in fact, two or more "cryptic species" and thus the species definition must be changed [14, 15]. Investigators can then reveal unknown properties of the entity under investigation.

In addition to distinguishing one thing from another, and rigorously identifying essential qualities – another function of scientific definitions is to provide cohesion to the conceptual schemes of a discipline [16]. Definitions provide this cohesion by accounting for all the *known elements of a system* – indicating where one system (and the discipline devoted to studying that system) ends and another begins. In this way, a scientific definition can reveal the existence of entities, whose parts cohere in a manner previously unrecognized. An example of this is the proposal that all living things form part of a singular "biosphere," a name given to all living things which, considered collectively, can exert on influence on planetary processes [17]. The properties of the elements of these previously unrecognized systems can then be carefully described, and its elements classified. To be clear: the biosphere as we currently understand it (see *Chapter 7*) has existed since the dawn of life, however it has only be named since the late nineteenth century. The identification of entities and the characterization of processes is the major task of ecology as a natural science.

Before a scientific investigation can get properly underway, it is necessary to identify and describe all the conceptual terms being employed in the analysis. Dictionaries and glossaries that circumscribe all concepts, entities, and processes, relevant to a discipline serve a useful function in undertaking any scientific work [18, 19].

To give a simple but concrete example of how a discipline proceeds with definitions of entities, processes, and concepts required for scientific work, consider an ecologist studying, let's say, a tropical stream (this example is drawn from the work of the great American ecologist Howard T. Odum [20]). In order to understand the system, all of the species relevant to the study must be identified. The inventoried species can be sorted into those functional categories most appropriate to ecosystem ecology, e.g. primary producers (photosynthetic organisms), consumers (herbivore, carnivores, etc.), and saprophytic species (those involved in the breakdown of dead organic matter). Thus, the ecologist is employing a definition of "species" (that is, assembling all genetically related individuals), and then these species are aggregated into categories defined by their ecological function. The array of conceptual terms used to define ecological processes are then defined. For example "net primary production" (that is, the process of accumulating energy or matter, minus the respiratory losses of photosynthetic or chemosynthetic organisms), "decomposition" (the breakdown of dead material), "nutrient cycling" (the cycles of key elements through the biological, geological, atmospheric, and hydrospheric compartments of the tropical system being investigated) will be defined, measured, and used to derive an understanding of the entire system [21]. The example reveals the role of scientific definitions both in making *distinctions* (identifying species) and in *combining* elements that might be otherwise considered distinct (functional categories).

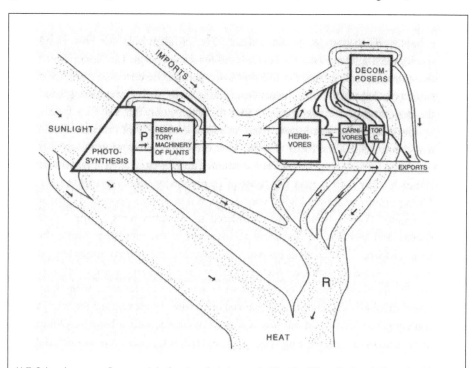

H.T. Odum's energy flow model of spring-fed stream in Florida (Silver Springs). H are herbivores, C are carnivores, and TC are top carnivores. Squares represent biotic pools [1] (This work has been released into the public domain by its author, Sholto Mau).

3.3 Coda

We use definitions in science to determine those conditions that are both necessary and sufficient to work with concepts. It may not always be possible to determine the essence of an entity, yet we will want to find the properties that identify an object in a unique way [22]. In the following section of the book, we identify a conceptual entity called a system, one that has proved almost uniquely useful in guiding environmental research.

References

1 Coates, P. (2013). *Nature: Western Attitudes since Ancient Times*. Wiley.

2 Morton, T. (2007). *Ecology Without Nature: Rethinking Environmental Aesthetics*. Harvard University Press.

3 Golley, F.B. (1998). *A Primer for Environmental Literacy*. New Haven, Connecticut; London: Yale University Press.

4 Oxford English Dictionary *Environment, n*. Oxford University Press (accessed January 2022).

5 Benner, S.A. (2010). Defining life. *Astrobiology* 10 (10): 1021–1030.

6 Gayon, J. (2010). Defining life: synthesis and conclusions. *Origins of Life and Evolution of the Biosphere* 40 (2): 231–244.

7 Baldwin, J.R., Faulkner, S.L., and Hecht, M.L. (2006). A moving target: The illusive definition of culture. In: *Redefining culture: Perspectives across the disciplines* (ed. J.R. Baldwin, S.L. Faulkner, M.L. Hecht and S.L. Lindsley), 3–26. Lawrence Erlbaum Associates.

8 Oderberg, D.S. (2011). Essence and properties. *Erkenntnis* 75 (1): 85–111.

9 Nakhnikian, G. (1971). *Elenctic Definitions*, 125–157. Palgrave Macmillan.

10 Antonelli, G.A. (1998). Definition. In: *Routledge Encyclopedia of Philosophy* (ed. E. Craig). Routledge. https://www.rep.routledge.com/articles/thematic/definition/v-1 (accessed 16 November 2022).

11 Belnap, N. (1993). On rigorous definitions. *Philosophical Studies: An International Journal for Philosophy in the Analytic Tradition* 72 (2/3): 115–146.

12 Correia, F. (2017). Real definitions. *Philosophical Issues* 27 (1): 52–73.

13 Janus, S.Q. (1940). The role of definition in psychology. *Psychological Review* 47 (2): 149.

14 Adams, M., Raadik, T.A., Burridge, C.P., and Georges, A. (2014). Global biodiversity assessment and hyper-cryptic species complexes: more than one species of elephant in the room? *Systematic Biology* 63 (4): 518–533.

15 Roca, A.L., Georgiadis, N., Pecon-Slattery, J., and O'Brien, S.J. (2001). Genetic evidence for two species of elephant in Africa. *Science* 293 (5534): 1473–1477.

16 Caws, P. (1959). The functions of definition in science. *Philosophy of Science* 26 (3): 201–228.

17 Vernadsky, V.I. (1945). The biosphere and the noosphere. *American Scientist* 33 (1): 1–12.

18 Perelet, R., Mason, P., Markandya, A., and Taylor, T. (2014). *Dictionary of Environmental Economics*. Routledge.

19 Allaby, M. and Park, C. (2013). *A Dictionary of Environment and Conservation*. Oxford University Press.

20 Odum, H.T. (1956). Primary production in flowing waters 1. *Limnology and Oceanography* 1 (2): 102–117.

21 Chapin, F.S. III, Matson, P.A., and Vitousek, P. (2011). *Principles of Terrestrial Ecosystem Ecology*. Springer Science & Business Media.

22 Burge, T. (1993). Concepts, definitions, and meaning. *Metaphilosophy* 24 (4): 309–325.

Section Two

The Concept of Systems

4

Everything Is Connected: The First Rule of Ecology

> In its essence, a system is a collection of parts that interact to produce a unity or whole. A machine provides a model for a system: it has several parts that interact in fairly predictable ways to produce the behavior of an integrated whole.
>
> *Source:* Kazimir Severinovich Malevich / Public Domain.

Environmental science is often described as a "systems science" [1, 2]. What is meant by this is that the phenomena relevant to environmental enquiry, the analysis of environmental problems, and the application of insights from these disciplines are founded on a particular class of models: systems models. These models were developed primarily for the purpose of facilitating the investigation of concerns found in common across several fields of research [3]. Because so many disciplines characterize themselves as systems endeavors – for instance, organismal biology, economics, some subdisciplines of psychology, anthropology, and other social sciences – the sources of system thinking are therefore exceedingly diverse. The precise inspiration for systems thinking in any one discipline can be difficult to identify [4, 5]. Despite the variegated nature of models available for systems analysis, the environmental sciences have drawn especially on the general systems theory (hereafter, GST), cybernetics, information theory, game theory, and system dynamic modeling. In this chapter, rather than providing a comprehensive account of systems thinking in all its manifestations (such accounts can be found elsewhere [5, 6]), we consider just those aspects of systems thought that may prove helpful in understanding the models discussed throughout the remainder of this book. This chapter introduces some of the basic concepts of systems thinking; the second chapter of the section examines a more complex set of system ideas especially relevant for the interrogation of human impacts.

4.1   What Is a System?

Do the objects on the writer's table form a system? *Source:* Liam Heneghan (Author).

By "system" in GST is meant "a collection of objects organized according to a unity or whole" [3]. A prominent source of systems thinking in the environmental sciences are ideas developed by the Austrian theoretical biologist Ludwig von Bertalanffy (1901–1972). His work on systems is summarized in his influential book *General System Theory: Foundations, Development, Applications* (1968) [3]. It is from von Bertalanffy's account of systems that we primarily draw this sketch.

Systems, from the perspective of GST, are everywhere, but not everything is a system. It is helpful to start by pointing out those things that *would not qualify* as a system. The table at which you sit to read this book may be strewn with a variety of objects. (If you are not at a desk, glance around for the nearest one!) On the desk at which I write this chapter, for example, there is a computer, a notebook, a wrist watch, a radio, a cup, a pencil sharpener, a paint box, and so on (as illustrated in photo on this page). We can refer to this assortment as "a collection of objects," but they cannot be said to be particularly organized. The notebook on the table contains random jottings for another writing project, the cup contains a few last sips of tea I made earlier in the morning, the radio is where I set it aside after listening to a show last evening. None of these objects has an immediately apparent functional relationship: they are strewn about the table based upon a whim after I last used them. This collection of objects would clearly not qualify as a system from the GST perspective *unless*, for example, in this analysis, I included myself as an entity in the system and you as an observer. The analysis would then become a "second-order" investigation, where you can ask the (admittedly uninteresting) question of how this author organizes his work life.

Though the constellations of objects on my desk do not constitute a system (in the absence of the author as organizer of the system), the computer, radio, cup, and watch are themselves clearly "organized" objects. Each of these items is comprised of component parts that are repeated in space and time – open up *any* radio and you will find a predictable suite of component parts. Furthermore, the components of each of these entities interact in a predictable manner and thus the parts unite to serve a particular function. Employing our definition of system as a collection of objects organized according to a unity or whole, a radio, a body, a city, an economy, a family, and a governmental institution, all qualify as systems.

Reflecting upon the variety of these systems, the diversity of ends they satisfy, and the heterogeneity of interactions that regulate them, it can be easy to appreciate why systems theory appeals to such a range of disciplines.

4.2 Everything Is Connected

Systems analysis directs our attention both to the entities and to the processes that connect them. Therefore, systems-level investigations permit us to understand how linked parts contribute to the behavior of a whole: for example, how do the interactions between individual humans constitute a family? Two forms of connection are especially important to us: *events* and *processes*, and we will distinguish these in more detail in the next section.

John Muir (1838–1914). Naturalist, wilderness advocate, and earlier campaigner in support of National Parks in the United States. Muir's writing has been especially influential in making the idea of "connection" a central one in ecological thought. *Source:* The Library of Congress

It is a truism of ecology that everything is connected – ecology is thus the quintessential systems science [7]. The notion of connection in natural systems is picturesquely captured by the Scottish-American naturalist John Muir (1838–1914) who wrote in his book *My First Summer in the Sierra* (1911): "When we try to pick out anything by itself we find it hitched to everything else in the universe" [8]. The thought of interconnection is also referred to, strikingly, by Charles Darwin as the "tangled bank." Darwin concluded *The Origin of Species* (1859) with this stimulating metaphor [9]: "It is interesting to contemplate a tangled bank, clothed with many plants of many kinds, with birds singing on the bushes, with various insects flitting about, and with worms crawling through the damp earth, and to reflect that these elaborately constructed forms, so different from each other, and dependent upon each other in so complex a manner, have all been produced by laws acting around us."

Elsewhere in *The Origin of Species,* Darwin provides a detailed example of the sort of organismal interconnections that shape the behavior of pastoral communities that surrounded his home in South East England. I will quote the section in full, so that I can draw upon the example for the remainder the chapter.

From experiments which I have tried, I have found that the visits of bees, if not indispensable, are at least highly beneficial to the fertilisation of our clovers; but humble-bees alone visit the common red clover (*Trifolium pratense*), as other bees cannot reach the nectar. Hence I have very little doubt, that if the whole genus of humble-bees became extinct or very rare in England, the heartsease and red clover would become very rare, or wholly disappear. The number of humble-bees in any district depends in a great degree on the number of field-mice, which destroy their combs and nests. . . Now the number of mice is largely dependent, as every one knows, on the number of cats; and Mr. Newman says, "Near villages and small towns I have found the nests of humble-bees more numerous than elsewhere, which I attribute to the number of cats that destroy the mice." Hence it is quite credible that the presence of a feline animal in large numbers in a district might determine, through the intervention first of mice and then of bees, the frequency of certain flowers in that district! [9]

The entities and connections between them that Darwin identifies here are various: humble-bees (now, more commonly, called bumblebees) visiting the flowers of local herbaceous plants, mice preying upon and destroying bumblebee nests; cats feeding upon mice, and, indeed, cats being tended to by humans. By elucidating certain ecological interactions and determining how these are lined up in sequence (which comes first, what happens next, what is the result?), patterns of nature that might otherwise have remained cryptic, or the analysis of events that seem only tenuously connected, are understood in their ecological significance. Darwin's example illustrates the fundamental task of the naturalist: on the one hand, to identify and catalog the most important components of the system and, on the other, to sift through those ecological interactions in order to determine how a system is organized as a unity or whole. Interestingly, although Darwin's example remains a good one, later research indicates that field mice are seldom observed raiding bumblebee nests and, in fact, the systems he described may be considerably more complex [10]. The same sort of connections may prevail, but the system components may be different from those conjectured by Darwin.

4.3   Events, Processes, and System Behavior

The behavior of a system such as the one described by Darwin is determined by the patterns of interactions between the component parts. By behavior is meant any discernable change of a system with respect to its surroundings [11]. Though this definition is vague, it is usefully so, because the definition can be applied to all systems from electrons, automobiles, ecosystems, and even the universe itself. In each case, changes to the system (compared to their surroundings) arise from the occurrence a set of events.

> An **event**, in system terms, connects system elements and exists only for the time interval over which it occurs. When events recur predictably in difference places and at different times, they are considered to be parts of *processes*.

It is helpful in analyzing a system to clearly distinguish objects or entities (the etymology of which is *ens*, that is, "a thing") from those events and processes that connect things [12]. Let us first consider the meaning and significance of the term "event." We could describe a single bumblebee visit to a single clover flower as an *event*; likewise, a mouse destroying a bumblebee nest is also an *event*. But what makes an event, an event? Furthermore, what distinguishes *an event* from *a process*? An event is something that exists solely over the interval in which it occurs. Each event may have several temporal parts – for example, the bumblebee visit may have its stages: the bee lands, it walks about, it takes in sensory cues, processes these cues, it feeds, and then it departs. Though an event has temporal components, it does not have parts that are truly spatial. The ambiguity of its spatial characteristics is what distinguishes events from the entities it connects, like bumblebees and flowers. Bumblebees and plants have properties that will endure even as they change; events, of course, may endure – a bumblebee visit can last for a few minutes – but they cannot be said to meaningfully change: an event in its totality is a singular occurrence [13].

Rare, or even one-off events, can be extremely important in determining the trajectory of a system. A general case for the importance of rare events in structuring systems is made by statistician and essayist Nassim Taleb in his book, *The Black Swan: The Impact of the Highly Improbable* (2007) [14]. Taleb's account of the importance of low probability but highly consequential events (i.e. a black swan) has been most often discussed in the financial world and in the social sciences. However, there is emerging evidence of their importance in ecology and evolutionary biology [15, 16]. Extreme weather events, disease outbreaks, and ecological invasions, for example, or such factors occurring in odd combinations can have enduring legacies for ecological systems.

A **process** is a continuous, regular (and often recurring) action performed in a definite sequence that influences system behavior.

Though attentiveness to rare events – especially ones that may have catastrophic consequences – is becoming more prevalent in the environmental sciences, more consideration is paid to those events that repeatedly and predictably reoccur. Of especial importance are those events that recur in regular sequence and that instigate predictable changes in the system. We can *call a set of events* that recur in a prescribed sequence and that results in a particular outcome a *process*. Processes have both a direction and an internal order and these emerge from interactions linked in a chain of cause and effect [17]. To return to Darwin's example, the bumblebees visitation to a plant will be seen a no mere collision that occurs once only – rather, the visit is an instance of the *process* of pollination (in which both plant and bumblebee benefit). Bumblebee nest invasion by mice (or other mammals) is an example of the *process* of predation (in which mice benefit, though bumblebees not at all).

The behavior of all systems is governed by many such processes. Upon initial inspection, certain events might seem unique to that system. Let us say that you notice a squirrel burying an acorn in a woodland. To an untrained eye this may seem like a one-off event. However, storing of seed (acorns, in well-studied cases) by mammals is very influential in woodland succession [18]. By burying the seed, this food resource can be reserved for consumption by the mammal in leaner times. Often, though, such stores are forgotten, and thence germination of these underground stores can markedly influence plant demography and successional change of ecosystem development [19]. Storage events such as these seem to be quite specific, but similar events occurs frequently in widely divergent ecological systems [18, 20–22]. Whenever it occurs, this process is referred to as "caching" (from the French *cacher* meaning to hide). Caching, we now know, is not confined to mammals hiding seed: birds also engage in ecologically significant caching of resources, and aspects of this process are comparable to the hiding of non-seed resources by other animals, including spiders [19, 23]. The conceptual net can, however, be spread even wider: caching processes have been analyzed in early human agricultural communities [24, 25], and the term is also employed in both communication and computer science [26, 27]. We might even describe the storing of comforting bars of chocolate about the home as caching (indeed, it is sometimes referred to as "squirreling away" treats). Caching can be said to be an instance of a general systems process.

The fairly narrowly focused observations above can allow us make a more general point. The study of similar processes in multiple disciplines sheds light on the general nature of system interactions. In facilitating interdisciplinary discussion, specialists can compare and contrast the mechanisms of interaction between the component parts of systems that are seemingly disparate. Thus, a crystallographer, a population biologist, a physiologist, and sociologist can learn from similarities in growth processes pertaining to their widely different empirical fields, while remaining attentive to the relevant differences.

It is, however, important to appreciate that although systems theory permits thinking across the disciplines, systems analysis does not replace a discipline. Rarely should anyone regard themselves as a systems theory specialist: rather, one should be able to draw upon the conceptual resources of systems thinking to develop a clearer and more interdisciplinary picture of the problem to which one is devoting energy [28].

So far, we have identified the following as important to system thinking: the need to identify all the parts that make up the components of entity – that is, the parts that constitute an integrated whole. These parts are connected by means of a variety of processes in a manner that determines the behavior of the whole system. From the perspective of systems theory many things may seem connected.

4.4   Everything Is Connected…but Some Things Are More Connected Than Others

The ecological thought of connection is a compelling one; it is commonly referred to as the first law of ecology [29]. That everything is connected to everything else is a powerful goad for investigating how even relatively modest disturbances (by humans or those occurring within the nonhuman sphere) might ramify throughout a complex web of interacting entities, creating surprising outcomes. It is the task of the environmental sciences to uncover those connections, and to do so with a view to averting harmful consequences for humans and the rest of nature.

An extreme, though instructive, instance of the idea that within a given system everything connects is exemplified by the "Butterfly Effect," a metaphor attributed to meteorologist Edward N. Lorenz [30]. "Does the flap of a butterfly's wings in Brazil set off a tornado in Texas?," is a question that Lorenz famously posed in a lecture in 1972 [31]. The Butterfly Effect holds out the slightly terrifying prospect of a world where certain capricious events can create chaos [32]. If that most poetically delicate of gestures – the wings of a butterfly flapping – can wreak havoc, then this is surely an unsafe world to live in. Of course, in reality, every time a butterfly flaps its wings, we need not, in fact, run for cover. Why not? It is because a large lepidopteran influence on meteorological systems is infinitesimally improbable. The world seems safe from the menace of butterfly-induced hurricanes. Lorenz's point was not that the earth systems are so mechanically rigged that even tiny events will reverberate in colossal ways; rather his point was that small changes in initiating conditions in any system can have unforeseen implications for the development of the system [31]. The point has implications for ecology, evolutionary studies, and the social sciences [31, 33–35].

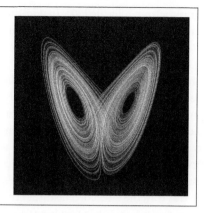

The butterfly effect as seen in the Lorenz system with parameters set to σ = 10, r = 28, b = 8 3. Though it is sometimes assumed that the name "Butterfly Effect" is derived from the shape of this plot, but that is not the case: the effect was named before this figure became widely used. *Source:* Wikimol / Wikimedia Commons / Public Domain.

This caution against extreme interpretations of the Butterfly Effect might allow us supplement the first law of ecology that "everything is connected" to *"everything is connected but some things are more connected than others."* Undestanding that this is the case is very important to appreciating the notion of *system stability* – that is, the ability of systems to endure in their habitual behavior despite the omnipresent of disturbances. Two aspects of general system functioning ensure that systems are generally stable in the face of small disruptions. On the one hand, many systems are modularly designed (like popular Swedish-made furniture, each component part is a relatively stable sub-unit), and on the other, though related to this, systems can have boundaries – either physical or dynamical ones – that contribute to durable functioning by ensuring that each disturbance doesn't ripple across the world. (These system features are explored in *Chapter 5*).

4.5   Connectivity, System Stability, and Resilience

Disturbance is ubiquitous in the natural world: hurricanes, forest fires, pest outbreaks, landslides, and floods operate on a landscape scale, but even in small, spatially confined arenas, disturbance can influence the lives of organisms. (Note that the flapping of insect wings is not included in this roster.) A brief drought can be experienced as a disruptive to creatures with short life spans. Thus, a disturbance event can vary both quantitatively (how large or small it is) and qualitatively (that is, there is a large variation in both the types of disruptive events and in how these are experienced by organisms).

There is no one definition of ecological disturbance that satisfies all environmental circumstances, or that applies at all ecological scales. However, the following definition from ecologist Edward J. Rykiel will serve for our purposes:

> [A disturbance is] a cause; a physical force, agent, or process, either abiotic or biotic, causing a perturbation (which includes stress) in an ecological component or system; relative to a specified reference state and system; defined by specific characteristics [36].

In addition to considering the properties of individual disruptive agents, the analysis of disruption (or perturbation, to use a common ecological synonym) should be assessed by evaluating impacts upon affected communities and ecosystems. Disturbance can impose an additional level of mortality on individuals (above background rates of mortality) within a community or can forcing emigration from a locale [37]. In addition to altering population density and community composition through mortality and emigration, disturbance can further result in altered ecosystem functioning (i.e. changes in those ecological processes that control, for example, fluxes in energy and nutrient exchange).

Ecologists have long been interested in evaluating the process of system recovery after a disruptive event subsides (events can, of course, vary greatly in their duration). A key concept in the study of the recovery of systems after an ecological disturbance is the notion of *stability*. Stability is a theoretical concept: an ecological system, in theory at least, may exist in a state of equilibrium (or balance) until a disruption occurs. A population is described as being in *stable equilibrium* if population size returns to an equilibrium value after a disturbance. If the population diverges from its original size following a small perturbation it is said to be in a state of unstable equilibrium. Such a state is often temporary. In this way, the effect of a disturbance can be measured against the supposed stability of a community or ecosystem. Ecologists have also devoted much attention to how quickly a system will return to equilibrium after a disturbance. The examination of this property – sometimes called "engineering resilience" – generated many interesting results in theoretical ecology [38].

More recently, ecologists have been interested in another system property: namely "ecological resilience" [39]. That is, under what circumstance can a system retain essential system features without transitioning to a new stable state? For example, when nitrogen or phosphorus is added to an aquatic system in uncustomary quantities, an ecologist might be interested if that body of water can continue to function without flipping to a new state (e.g. a clear lake can up to a point resist a transition to a highly polluted and turbid condition).

Resilience, as a property of ecological systems, has emerged as a key concept in contemporary management [40]. It can be defined as follows:

> Resilience is the capacity of a system to absorb disturbance and reorganize while undergoing change so as to still retain essentially the same function, structure, identity, and feedbacks [41].

Key aspects of the emerging conceptual frameworks concerning "resilience thinking" include:

1) The occurrence of multiple states for many ecosystems (for example, a region can be vegetated or a desert; a tropical forest can under some conditions become a scrubland).
2) Systems up to a certain threshold can to absorb change in driving variables (for example, alterations in moisture, temperature, and soil nutrient levels).
3) The transition from one relatively stable state can occur rapidly and discontinuously (that is, there may not be a graduated transitional state).

Concepts of disturbance and resilience are applied not only to purely ecological systems, but also are relevant to the study of social-ecological systems. Historian of ecology Donald Worster remarked upon this development in his important overview of the development of environmental science, *Nature's Economy: A History of Ecological Ideas* (1977):

> [S]cientists have abandoned [the] equilibrium view of nature and invented a new one that looks remarkably like the human sphere in which we live. We can no longer maintain that either nature or society is a stable entity. All history has become a record of disturbance and that disturbance comes from both cultural and natural agents, including droughts, earthquakes, pests, viruses, corporate takeovers, loss of markets, new technologies, increasing crime, new federal laws, and even the invasion of America by French literary theory [42].

It is not clear from the above passage that Worster regarded this preoccupation by ecologists with disturbance as, unambiguously, a good thing. However, it undoubtedly underscores that idea that a systems framework can connect our thinking about the complexity of nature with an examination of social systems [43].

4.6 Coda

This chapter served to introduce you to the concept of a system – this manner of framing problems has far-reaching implications for many disciplines. In some of its uses, system thinking can seem restrictive – by placing objects into seemingly inflexible boxes – though this approach can be helpful for the analysis of mechanical and engineered systems. However, in environmental disciplines, system thinking is often employed to elucidate complex natural entities that are comprised of many component parts, and may often lead to insights about the surprising properties of such entities that defy reductive thinking. These complex environmental systems and their properties are the subject of the next chapter.

References

1 Cunningham, W. and Cunningham, M.A. (2010). *Principles of Environmental Science*. McGraw-Hill Higher Education.

2 McKinney, M.L. and Schoch, R.M. (2003). *Environmental Science: Systems and Solutions*. Jones & Bartlett Learning.

3 Von Bertalanffy, L. (1969). *General System Theory: Foundations, Development, Applications*. New York: George Braziller.

4 Hammond, D. (2002). Exploring the genealogy of systems thinking. *Systems Research and Behavioral Science* 19 (5): 429–439.

5 Hammond, D. (2010). *The Science of Synthesis: Exploring the Social Implications of General Systems Theory*. University Press of Colorado.

6 Von Bertalanffy, L. (1972). The history and status of general systems theory. *Academy of Management Journal* 15 (4): 407–426.

7 Morton, T. (2010). *The Ecological Thought*. Harvard University Press.

8 Muir, J. (1911). *My First Summer in the Sierra*. Houghton Mifflin.

9 Darwin, C. (1996). *The Origin of Species: Oxford World's Classics*. Oxford University Press.

10 Goulson, D., O'Connor, S., and Park, K.J. (2018). The impacts of predators and parasites on wild bumblebee colonies. *Ecological Entomology* 43 (2): 168–181.

11 Rosenblueth, A., Wiener, N., and Bigelow, J. (1943). Behavior, purpose and teleology. *Philosophy of Science* 10 (1): 18–24.

12 Salthe, S.N. (1985). *Evolving Hierarchical Systems*. Columbia University Press.

13 Galton, A. and Mizoguchi, R. (2009). The water falls but the waterfall does not fall: new perspectives on objects, processes and events. *Applied Ontology* 4 (2): 71–107.

14 Taleb, N.N. (2007). *The Black Swan: The Impact of the Highly Improbable*. Random House.

15 Anderson, S.C. et al. (2017). Black-swan events in animal populations. *Proceedings of the National Academy of Sciences* 114 (12): 3252–3257.

16 Logares, R. and Nuñez, M. (2012). Black swans in ecology and evolution: the importance of improbable but highly influential events. *Ideas in Ecology and Evolution* 5: 16–21.

17 Emmet, D. (1998). Processes. In: *Routledge Encyclopedia of Philosophy* (ed. E. Craig). Routledge. https://doi.org/10.4324/9780415249126-N047-1.

18 Steele, M.A. et al. (2001). Cache management by small mammals: experimental evidence for the significance of acorn-embryo excision. *Journal of Mammalogy* 82 (1): 35–42.

19 Vander Wall, S.B. (1993). A model of caching depth: implications for scatter hoarders and plant dispersal. *The American Naturalist* 141 (2): 217–232.

20 Darley-Hill, S. and Johnson, W.C. (1981). Acorn dispersal by the blue jay (Cyanocitta cristata). *Oecologia* 50 (2): 231–232.

21 Longland, W.S. and Clements, C. (1995). Use of fluorescent pigments in studies of seed caching by rodents. *Journal of Mammalogy* 76 (4): 1260–1266.

22 Abbott, H.G. and Quink, T.F. (1970). Ecology of eastern white pine seed caches made by small forest mammals. *Ecology* 51 (2): 271–278.

23 Smith, C. and Reichman, O. (1984). The evolution of food caching by birds and mammals. *Annual Review of Ecology and Systematics* 15 (1): 329–351.

24 Cunningham, P. (2011). Caching your savings: the use of small-scale storage in European prehistory. *Journal of Anthropological Archaeology* 30 (2): 135–144.

25 Hurst, S. (2006). An analysis of variation in caching behavior. *Lithic Technology* 31 (2): 101–126.

26 Karlin, A.R. et al. (1988). Competitive snoopy caching. *Algorithmica* 3 (1): 79–119.

27 Hsu, H. and Chen, K.-C. (2015). A resource allocation perspective on caching to achieve low latency. *IEEE Communications Letters* 20 (1): 145–148.

28 Boulding, K.E. (1956). General systems theory—the skeleton of science. *Management Science* 2 (3): 197–208.

29 Commoner, B. (2020). *The Closing Circle: Nature, Man, and Technology*. Courier Dover Publications.

30 Lorenz, E.N. (1995). *The Essence of Chaos*. University of Washington Press.

31 Hilborn, R.C. (2004). Sea gulls, butterflies, and grasshoppers: a brief history of the butterfly effect in nonlinear dynamics. *American Journal of Physics* 72 (4): 425–427.

32 Gleick, J. (2008). *Chaos: Making a New Science*. Penguin.

33 Zhen, W. et al. (2020). Specific "Unlocking" of a Nanozyme-based butterfly effect to break the evolutionary fitness of chaotic tumors. *Angewandte Chemie* 132 (24): 9578–9584.

34 Vernon, J.L. (2017). Understanding the butterfly effect. *American Scientist* 105 (3): 130.

35 Adlard, R.D., Miller, T.L., and Smit, N.J. (2015). The butterfly effect: parasite diversity, environment, and emerging disease in aquatic wildlife. *Trends in Parasitology* 31 (4): 160–166.

36 Rykiel, E.J. (1985). Towards a definition of ecological disturbance. *Australian Journal of Ecology* 10 (3): 361–365.

37 Bender, E.A., Case, T.J., and Gilpin, M.E. (1984). Perturbation experiments in community ecology: theory and practice. *Ecology* 65 (1): 1–13.

38 May, R.M. (2019). *Stability and Complexity in Model Ecosystems*. Princeton University Press.

39 Hastings, A. and Gross, L. (2012). *Encyclopedia of Theoretical Ecology*. University of California Press.

40 Walker, B. and Salt, D. (2012). *Resilience Thinking: Sustaining Ecosystems and People in a Changing World*. Island Press.

41 Walker, B. et al. (2004). Resilience, adaptability and transformability in social–ecological systems. *Ecology and Society* 9 (2).

42 Worster, D. (1994). *Nature's Economy: A History of Ecological Ideas*. Cambridge University Press.

43 Scheffer, M. (2020). *Critical Transitions in Nature and Society*. Princeton University Press.

5

Complex Environmental Systems

Flocks of birds are often studied as examples of self-organizing systems – the properties of the flock are not readily predicted from the behavior of individual birds. How does the behavior of the flock emerge from its constituent parts? Can you think of other examples where the behavior of a collective seems to be "greater than the sum of its parts?"

Plate 67 Red-winged Starling or Marsh Blackbird. John James Audubon. Series: *Birds of America* 1827–1838

The previous chapter introduced some of the more basic concepts of systems theory. As we saw in that chapter, systems are entities comprised of heterogeneous parts that constitute an organized whole: a living cell, a body, a car, and the solar system are all examples of systems. The interaction of system parts both by means of recurring processes and by means of rare or even one-off events determine the behavior of the integrated system whole. Analysis of the phenomena of both the natural and the social worlds can reveal how such connections between the elements of a system – some of these connections often being quite cryptic – can determine how the behavior of a given phenomenon is manifested. We have illustrated this principle in the previous chapter with Charles Darwin's observations that the depredations of prowling cats – mediated by a several intervening, and somewhat surprising, steps – can determine the population density of wild flowers.

Systems theory develops a set of unified principles applicable to a wide variety of circumstances and therefore, potentially, this body of theory can help solve problems in an array in disciplines. Beyond the relevance of systems theory *within* each discrete discipline, because of its generality, systems theory has facilitated the development of a cross-disciplinary understanding of certain

A Primer on Human Impacts on the Environment: The Conceptual Approach, First Edition. Liam Heneghan.
© 2023 John Wiley & Sons Ltd. Published 2023 by John Wiley & Sons Ltd.

phenomena. Complex problems that inherently require insight from several disciplines are especially assisted by adopting a systems perspective. This situation certainly pertains in the case of many global environment issues: anthropogenic climate change, for example, has physical, biological, and social implications. Its study, therefore, cannot be the subject of any one discipline.

Climate is the weather pattern of a region considered over a period of time measured in decades (30 years being the typical period). Climate can and does change in natural circumstances; however, *anthropogenic climate change* refers to *accelerated* climatic changes since the 1800s under the influence of human activities (driven primarily by burning fossil fuels like coal, oil, and gas). Systems analysis is crucial in linking changes in emissions occurring at local scales to alterations of climate at a planetary scale.

In what follows, I introduce the notion of a complex adaptive system (CAS) – the name given to a distinctive type of system that, under certain selective pressures, changes over time and that constitutes familiar ecological systems. I demonstrate the significance of system boundaries that give shape or definition to objects in the familiar world of nature. System boundaries also allow for interactions among seemingly separate parts of entities that are themselves aggregated into the larger wholes that constitute our clearly modular universe. A set of rules governing the relation of wholes and parts are then discussed. Throughout this chapter frequent references are made to the higher and lower levels in system hierarchies: at all times when considering these points it will be useful to have in mind a concrete example, such as a forest which can be considered a higher level when contrasted with the plants, animals, and microbes contained within that forest and which therefore occupy a lower system level.

5.1 Introducing Complex Adaptive Systems

Biological and environmental systems are often referred to as Complex Adaptive Systems (CASs). A CAS is usually spatially discrete (that is, they occur in a particular place). A given CAS is characterized as having various parts that are predictably repeated – for example, each tree has roughly the same *type* of parts (and, at a larger more inclusive scale, this pattern of repetition applies to an entire forested ecosystem). Every separate manifestation of a system is, of course, idiosyncratic (the trees in one oak forest, for example, are obviously not the same individual oaks that are found in another forest, and the precise number of individual trees will vary from place to place), but the presence of particular parts is generally predictably (oaks, for example, will, by definition, dominate in an oak forest). A CAS is self-organizing: the patterns detectable at the level of the whole (e.g. a given woodland ecosystem) emerge from interactions among the lower-level components of the system (i.e. from interactions among the plant, animal, and microbial communities operating within the woodland) though these patterns can be constrained by regional conditions (e.g. the climate). Statements about this tension between changing the assemblage of parts of a system and the capacity of the whole to absorb these changes will recur in what follows and is a point of central concern in complex systems theory.

The properties of the whole – sometimes the term "global level" is employed rather than the "whole" – cannot be predicted purely from a knowledge of the behavior of the individual elements from which it is constituted. The global properties are said to *emerge* from the interactions of the parts [1]. One would not readily predict, for example, the characteristics of the living brain from the properties of its dominant parts (the individual neurons and glia cells), nor is the behavior of a flock of birds reducible to knowledge about each and every bird. These principles of self-organization and emergence are applied to ecological systems, but can also be seen in to robotic, and other abiotic systems [2].

Ecological succession, an example of a system level internal selection process, refers to a somewhat orderly sequence of vegetation development that occurs when new substrates become exposed (after the cooling of a lava flow, for example), or as a biological community recovers after a major disturbance.

To the more general account of systems, CAS analysis adds the additional complexity of considering the role that *internal selective processes* play in changing a system's behavior over time [3]. By using the term "selection" here, what is intended are those sorting processes – learning, or adaptation, for example – that enhance or otherwise influence the replication of a system based upon the success of past interactions between that system and its environment. Selective processes can impact both the continuity and change of properties of a system; such processes determine which elements endure despite developmental change from iteration to iteration of the system over time. Additionally, selection can impact a system's development and reproduction. Because of selection, the most successful versions of an adaptive system are preserved and sustained through time.

Environmental systems are exemplary CASs: they have a reasonably consistent structure – that is, the diverse parts that constitute each system are maintained over time, and both their continuity and change are regulated by means of the variety of selective processes alluded to above [4]. Relevant selective processes for environmental systems include Darwinian natural selection over the long term, but also ecological sorting during the successional change of vegetation over medium-term time frames [5, 6]. CAS analysis can be applied to any environmental system (including social-ecological ones) with a capacity to adapt or learn: examples include sustainable development, global economic development, and climate adaptation [7].

Equipped with the account of CASs briefly sketched above, we will proceed in this chapter to elucidate those aspects of complex systems theory most relevant for understanding the composition of environmental entities and which can help in deriving solutions to environmental problems. In the following sections we place an emphasis on the following:

1) The *role of boundaries* in maintaining the spatial discreteness of systems and in determining their interactions with other (often adjacent) systems. The presence of system boundaries (both physical and dynamic ones) that serve to regulate connections within and between parts are crucial to the smooth functioning of modular parts in the totality of a system.
2) Understanding both the *modular and hierarchical nature* of ecological systems.
3) Evaluating the role of *processes* that regulate the composition of systems: their *parts* and *wholes*.
4) Introducing the concept of *emergent properties:* the circumstances in which the whole can be more (or occasionally less) than the sum of its parts.

5.2 System Boundaries

The word boundary derives from the Anglo-Latin word *bunda*, which signifies an external limit, that which circumscribes a thing. Part of the task of systems analysis is to determine the "bounds" of systems: that is, where one system ends and another begins. From the perspective of environmental systems analysis, two things about boundaries become immediately clear: firstly, boundaries are not always immediately apparent and, therefore, boundary setting is often the object of research, and, secondly, the close interrogation of boundaries can reveal the role they play in regulating system function.

> An **ecotone** is a zone of transition, or boundary, between adjacent ecological systems. There is great interest among ecologists in these marginal habitats because they often harbor a range of species from both systems in interesting combinations. As you leave one system and approach the neighboring systems, you can observe a rapid turnover of the species characteristic of each. Because of this, ecotones can be rich in species – a phenomenon referred to as the *edge effect*. In addition to providing habitat for individual organisms from each system, there are often species that specialize and are therefore of conservation interest [8].

The boundaries of a system can be quite vague. Darwin's example, considered in the last chapter, illustrated that in a pastoral system involving connections between cats, mice and bumblebee populations and that resulted in a manifestation of wildflowers at a particular density, the boundary of the system is somewhat diffuse. Other boundaries in biological and ecological systems, in contrast, can be sharp. For example, the wall of a cell, or the skin of body, are quite definitive. The margins of many ecosystems can be intermediate, with a zone of transition of indeterminate size – called an *ecotone* – that often exists between two habitats. Nonetheless, in this case, it usually becomes clear when you have left the meadow and entered the forest. Likewise, you will know you have left a terrestrial habitat and entered the aquatic realm when the water reaches your ankles.

Whether or not the boundaries are crisp or vague, all systems ultimately remain open to interactions with other systems. Thus, parts interacts to produce a given whole, and this whole (say, an individual organism bounded by its skin) can, in its turn, interact with other such wholes, and so on all the way up to the limits of the Earth as a world system. Even the global ecosystem itself is ultimately open despite the seemingly crisp distinction between the rind of the earth and space. The ecological systems of earth are animated by flows of energy from the sun, and the planet is materially subject to remote extraplanetary influences, even though these are subtle (approximately 5200 tonnes of micrometeorites fall to Earth every year, some of which can play ecologically relevant roles [9]). The fact that systems have boundaries and yet remain somewhat connected to other systems is the ultimate meaning of the amended first law of ecology: *everything is connected* (that is, boundaries between entities are porous) but *some things are more connected than others* (boundaries can serve to regulate and limit interactions between distant systems).

> Though *social-ecological systems* (SESs) are not new (as systems they are as old as the association between humans and nature) it is, nonetheless, a new term for the analysis of interactions between biophysical and social factors. The wise management of integrated SESs will be crucial in remediating environmental problems in the coming century.

What is the function of the numerous dry walls in the west of Ireland? What do the keep in; what do they keep out? A dry wall in the Burren, Co Galway, Ireland. *Source:* Michael Clarke Stuff / Wikimedia Commons / CC BY-SA 2.0.

It will be helpful to start with an example that illustrates both the multifunctional nature of system boundaries, but also reveals how their analysis can be productive of insights. This example comes from an evaluation of one particular landscape that is relevant to the human use of ecological resources – that is, it is a social-ecological example.

Parts of the landscape of the West of Ireland are characterized by noticeably small fields, poor soils, and innumerable drystone walls. Four hundred thousand kilometers of such walls traverse the Irish landscape [10]. Field enclosures bounded by walls can serve many functions: they are used for storage, to hold livestock and regulate their grazing, as well as providing shelter against the elements [11, 12]. The walls also demarcate patterns of ownership (one family's holding can be scattered across well separated fields) [12, 13]. The presence of walls can have positive implications for the local diversity of organisms; walls and associated fields may be classified as "novel cultural ecosystems" – that is, pleasing systems that have been shaped by human management, but which often retain elements that are rare and wild [14, 15]. All told, the walls are crucial to the genesis, maintenance, and functioning of these environmental systems.

In regular excursions with students to this part of Ireland, I ask them to reflect upon drystone walls and their role in the SESs of the West of Ireland. It is evident to them that the walls and enclosures receive some contemporary attention and maintenance, but most students also recognize that the walls have a longer archeological history. Indeed, they are often of great antiquity, dating at least to the Irish Bronze Age, circa 2500 BCE [16]. The cause of their emergence, in large part, resulted from the clearing of post-glacial debris by farmers [11]. Rocks dumped from the receding glaciers of the last Ice Age were aggregated into mounds; the walls developed merely as a convenient way of managing a rocky landscape. The cleared land and the fortuitously produced walls could then be committed to other needs [17].

It is clear – as my students recognized – that most of the current uses of these fields circumscribed by the drystone walls are a *consequence* of the walls becoming available. That is, the walls were not placed there *in order* to accommodate the function that they now serve [18]. This example – and my students' response to it – illustrates a general point about boundaries: they exist in various forms and for various reasons. Not only do the walls provide containers for entities (small grassy fields, cattle, an occasional abandoned tractor), but by defining and concentrating resources *within* boundaries, and by delimiting the extent of interactions *between* adjacent enclosures, their existence facilitates an innovation in function. Boundaries and walls are therefore, unsurprisingly, of central concern in any discussion of systems theory. Walls make certain things happen.

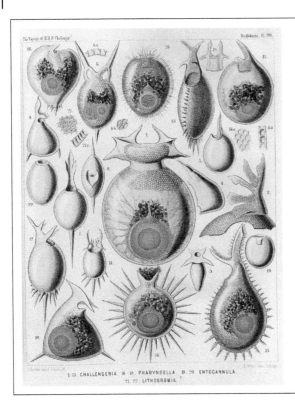

Radiolarians are single-celled organisms with intricate mineral skeletons. Within the confines of their cells wall, they have extraordinarily intricate structures. Cell walls serve both to contain, but also to facilitate selective exchange with the environment. *Source: John Murray / Wikimedia Commons / Public Domain.*

That boundaries serve to cradle the activities they enclose, as well as to regulate (actively or passively) communication between system subcomponents, is a commonplace observation regarding the CASs of biological and ecological systems: cells have their walls, plant ovules their integuments, chick embryos have germ walls, the heart its pericardium, plants and animals are enveloped in an epidermis, seeds and some invertebrates are contained within their testa; then there's the chorion of the insect egg, the sclera of the eye, the rind and peel of fruit, the exoskeletons of arthropods, the bark of trees, skin (yours and mine), the carapaces of beetles, the edges of forests, the reproductive isolation of species, and the geographic barriers between some major biomes. The living world is replete with boundaries: they surround things, facilitating exchange, and, as we have seen with drystone walls, incubate novel functions.

Once again, we can turn to Charles Darwin as a source of inspiration for systems thinking: in this case his example illustrates the importance of boundaries in facilitating innovation. Indeed, this insight of Darwin's concerning the role of boundaries might be one of the more singular thoughts in biology, since it concerns the origin of life itself. According to Darwin's often discussed thesis, life started with the establishment of a boundary. In a letter Darwin wrote to botanist J. D. Hooker in February 1871 he remarked:

> It is often said that all the conditions for the first production of a living organism are now present, which could ever have been present.— But if (& oh what a big if) we could conceive in some warm little pond with all sorts of ammonia & phosphoric salts,—light, heat, electricity &c present, that a protein compound was chemically formed, ready to undergo still more complex changes... [19]

Barriers function to seal in entities, but in dynamic systems they also serve to regulate flow into and out of a given system. After reading this section of the book, I invite you to set it down, and go seek out a set of doors, boundary markers, walls, gates, fences, and even valves: what functions do each perform? How do these barriers facilitate interaction between one side and what lies beyond? *Source:* Wellcome Images / CC BY 4.0.

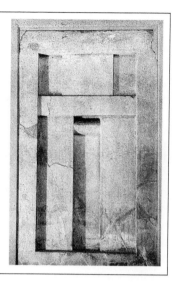

The idea that grounds Darwin's celebrated "warm little pond" thesis on the origin of life is simple enough. In order for the presumably fragile progenitors of life to stabilize and later propagate, they needed protection from the general tendency for concentrated activities to dissipate. A small warm pool, or as is the case in the relatively recent resurrection of Darwin's idea, "an inland geothermal fields within ponds of condensed and cooled geothermal vapor" may have provided the first protective boundary for life [20, 21]. Such a pond (geothermal or not) may be open to the world but it serves to concentrate important innovations – the genesis of life being the extremely important innovation in that case.

From the preceding examples that we have discussed we have established some important facts about system boundaries.

1) Boundaries are a biological and ecological commonplace (though they are also omnipresent in abiotic systems). In order for any entity to be investigated as an independent system, it *must* have boundaries.

2) These boundaries can be spatially sharp – a crisp edge can delineate a limit – or the boundary can be somewhat more vague (vague boundaries are sometimes referred to as "dynamic" ones). However, according to von Bertalanffy, an originator of General Systems Theory, vague boundaries exist only in "naive observation" (for example, the influence of your body seems clearly delineated by your skin, and yet the limit of your body as measured by, say, temperature, is dynamic rather than sharp, since all of us slightly heat the world surrounding us). All boundaries, in other words, are ultimately dynamic (that is, vague) [22].

3) The emergence of boundaries can *simultaneously* facilitate a wide range of system functions; oftentimes boundaries predate the functions to which they now contribute.

4) To facilitate innovation in system behavior, boundaries limit as well as facilitate communication between an ecological entity and its external environment.

The word **modular** derives from the Latin *modulus* meaning "a small measure." The dominant sense in which the term is used in systems thinking serves to denote an entity is constructed from component parts (especially when those components are similar if not identical). A theme of modularity runs through Western thought from at least the early third century BCE, when Epicurean atomism distinguished between physically indivisible particles ("atoms") and indivisible magnitudes.

Boundaries are ubiquitous and important, and yet we might still ask why, given a universe of all conceivable worlds, this is one in which *modularity* (and the boundaries upon which modularity depends) prevails? That is, why are entities of the natural world, more often than not, comprised of somewhat independent parts that dynamically interact at a variety of spatial scales? This is certainly the case for many entities of greatest interest to the sciences (from quanta to organisms) [23]. Not only is modularity conspicuous, these modular parts are found in hierarchally organized systems. The next section shows from a systems perspective why this might be so.

5.3   A Modular Hierarchical Universe

A typical CAS can be said to be a hierarchical system in which interaction among the component part (modules) contributes to the behavior of the whole. Every whole, in turn, can serve as a subcomponent in a progressively larger sequence of wholes (for example, atoms are parts of cells, cells are parts of tissues, tissues are parts of organs, and so on to the limits of the earth and beyond). The analysis of these complex hierarchical wholes reveals that an essential aspect of such systems is their *nonlinearity*, that is, the way that small changes in the composition of the parts can have large implications for the behavior of a system. Thus, system behavior depends upon but is not easily predicted from a knowledge of the properties of its parts. Moreover, the development of a system is *historically contingent* – thus, if the development of a system were to be replayed, the historical unfolding of that system might take an entirely different course (this is a possibility used for dramatic effect in the fictional genre of *alternate history*). Multiple possible outcomes exist for any system, each outcome being based upon the set of dynamic interactions that constitute it. That being said, if the same very strict selection criteria are applied each time, the system may indeed redevelop in a similar way after a disturbance (a hurricane, for example) [24–26]. Once in a while history can repeat itself.

[Los cinco libros ôl elfoxçabo z inuencible caualiero Tirante el blanco de roca falaba: Cauallero dela Barrotera. El qual por fu alta cauallería alcáço a fer príncipe y celar del imperio de grecia.]

The fictional genre of Alternate History is one where an outcome of a historical event is reimagined. Often some seemingly small change in the outcome of an events is imagined as altering the entire course of subsequent history.

In an early classic of the genre, Tirant lo Blanc our eponymous hero fights against the invaders threatening Constantinople and saves the Empire from destruction, contrary to the history of the region.

Speculate based upon your own life, or from your understanding of the ecological region in which you live, what would happen if history were to be repeated with small differences in initial conditions. *Source:* Centro Biblioteca Virtual Miguel de Cervantes / Wikimedia Commons / Public Domain.

It *is*, of course, possible that our perception of the world is "thoroughly contaminated" by a conceptual (or ideological) commitment to modularity and to an atomistic worldview. That is, it is possible that the perceived modularity of the world is simply a reflection of modularity in our cognitive architecture. If this is the case, we may be simply projecting onto the world some aspects of the organization of our own minds. Setting aside this troubling philosophical problem, the resolution of which is beyond the scope of this book, we should nevertheless look for an explanation of why many of the entities of interest to us in environmental science (indeed in most sciences) are in fact modular – that is comprised of parts within wholes [27].

An oft-repeated fable that offers some insight into why our world is modular originated with economist Herbert Simon. The fable concerns two watchmakers named Hora and Tempus [28–30]. Both craftsmen assemble watches from 1000 parts. Hora assembles his watches ploddingly, one component at a time. If he is disturbed while working, the watch fall to pieces and he has to start over. Tempus, on the other hand, proceeds by creating subassemblies (or modules) of 10 parts each; these are then aggregated into larger subassemblies of 100 parts. Eventually 10 of these subassemblies are used to finalize the completed watch. If Tempus is disturbed, he loses only *a small fraction* of the whole watch. If the frequency of disturbance is very high then Hora will struggle to complete a watch, whereas Tempus will continue to make splendid progress.

A take-home message from this parable is that in a world where entities are buffeted by disturbances that recur at periodic intervals – for example, by hurricanes, fires, or even telephone calls – those systems that are comprised of stable submodules seem to be more durable. To rephrase this, we can say that there is a "connection cost" associated with nonmodular systems, and, therefore, complex adaptive systems, such as a living organism, are selected for the possession of modularity, a condition which regulates the density of connections. Modularity, thus, contributes to "evolvability," that is, to the capacity of entities to adapt rapidly to new environments [31]. This propensity of modular organisms to evolve quickly has been demonstrated in computer simulations, and may have implications for other disciplines: neuroscience, genetics, and engineering. Searching for modularity of design is a purpose for which the tools of system theory are ideally suited.

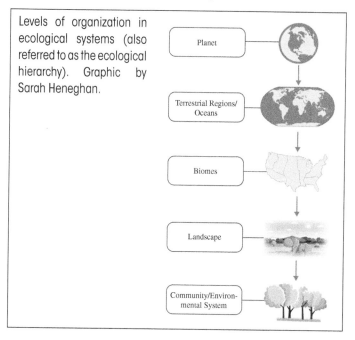

Levels of organization in ecological systems (also referred to as the ecological hierarchy). Graphic by Sarah Heneghan.

Planet

Terrestrial Regions/ Oceans

Biomes

Landscape

Community/Environmental System

Whatever the ultimate cause of the "lumpiness" of the world, it is undeniable that biological systems are modular and the greatest frequently of interactions occur within each of the clustered parts. However, this is not to say that no interaction occurs *between* these clustered nodes. In fact, it is by means of the interactions between parts (that are themselves comprised of parts) that the levels of biological organization emerges. Such levels of organization include, for example, aggregations such as cells, tissues, organs, organisms, populations, and ecosystems [32]. Each new whole (comprised of its component parts) is designated as the *upper level* as it comes into view and the analysis can then proceed to increasingly larger scales. Systems that are constituted by parts that are themselves comprised of parts, and that can then serve as constituents in even larger parts are said to be *hierarchical systems*. One tool for the study of such systems is called "hierarchy theory." Its key issues are "levels of organization" and the investigation of "problems of scale" (that is, the examination of novel phenomena as they emerge when the analytical lens gets more extensive both spatially and temporally) [33, 34].

Certain significant generalities regarding levels of organization can be immediately identified (adapted from [35]):

1) Each level of a system integrates the levels below and is associated with its own set of properties.
2) The behavior of an entity at a higher level is therefore dependent upon its lower levels.
3) Complexity increases as one ascends the levels of the hierarchy.
4) The lower levels of a system can be constrained by the levels above: for example, the cells of the body are not free to act independently of the body in which it finds itself (when they do act alone, this can devastate a system, for example, when the body responds to the proliferation of cancer cells).
5) System analysis can often be best undertaken by examining three levels: *the focal level* (that is, the one that is the focus of a given enquiry), *one level below the focal level* (it is from this level that we often seek an understanding of the mechanisms regulating the focal one), and *the level above the focal level* (this is the level that imposes constraints upon the focal level).
6) A disturbance at any level in the organization of the system can reverberate throughout the whole (though, often, the whole may be able to dampen the signals from lower levels and resist change). Changes at any level in the right circumstances can provoke nonlinear modifications in the system: a small in change can have a large effect if it occurs in the right circumstances.

5.4 Holons and Disciplinary Specialties

Each level in an ecological hierarchy can and often is studied independently of the others. Author, and systems theorist, Arthur Koestler uses the term *holon* to designate a hierarchical level in any system that can, and usually is, studied independently of others, even though the level is also part of a greater whole [29, 30].

> A **holon** is a level in a hierarchically organized system that can be studied as both an autonomous whole, but also as a dependent part of a larger system.

The behavior of each level in the *ecological hierarchy* (i.e. each holon) is governed by its own set of principles. Thus, the study of the dynamics of the individual organisms of a single species occupying the same locale at a given time is called *population ecology*. The types of problems investigated in population ecology include, for example, an exploration of the consequences of exponential and other growth forms (which are the result of reproduction), the self-limitation of populations (including questions about how populations are regulated), population oscillations (questions concerning population booms and busts), life history strategies (questions concerning birth rates, death rates, and other evolved demographic characteristics of each species), and the spatial organization of populations [36]. Populations drawn from *several species* together in the same space, simultaneously, form higher-level holons called *ecological communities* and the regularities governing communities form their own subdiscipline in ecology. The problems investigated by community ecologists include inter-specific competition, predation, mutualisms, food web stability, and the principles of biodiversity. When all the community of organisms in a region are considered in their relationship with the abiotic world, this holon is referred to as an *ecosystem*. Ecosystem ecologists study the flows of energy and the cycling of nutrients through and within this holon [37, 38]. Though each ecologist may identify as a specialist at the level of the holon of population, community or ecosystem, rarely are ecological patterns at any one level entirely examined without references to another holons [39].

An important aspect of studying holons is therefore that each is stable enough to bear disciplinary inspection, but cannot be fully understood outside of the hierarchical context in which they occur. This is precisely what Arthur Koestler meant when he wrote:

> The holons which constitute an organismic or social hierarchy are Janus-faced entities: facing upward, toward the apex, they function as dependent parts of a larger whole; facing downward, as autonomous wholes in their own right [29].

5.5 The Regulation of Parts and Wholes

Another aspect of the regulation of hierarchical systems worth elucidating are the logical rules that pertain to the mutual dependence of all hierarchical levels. Intuitively, for example, we can recognize that if all individuals of all populations of all the species in a community drop to zero (that is, become extinct) there is no longer any community to study. As a general rule the number of elements in a lower hierarchical level is greater than there are instances of the holon under interrogation: thus, there are fewer communities than there are populations of species, fewer cities than neighborhoods, and fewer organisms than the cells from which each organism is constituted. Upper levels in a hierarchy survive as long as some of the constituent parts persist.

The upper-level of a hierarchy in its turn, can serve to constrain lower levels. For example, the abiotic conditions of an ecosystem – the regional air temperatures and soil fertility levels – can influence the struggle for resources among species, and that struggle in turn can affect the life history of organisms. In this way, a higher-level holon can set limits on the behavior of the system parts beneath it.

As we have established, lower-level holons are more numerous, and smaller, than upper level ones. The dynamics that are characteristic of lower levels tend therefore to occur on smaller spatial scales than those higher in the hierarchy, and the processes associated with this lower level will occur more rapidly. Even though occupancy of lower levels in a system, as we have mentioned, is required for the survival of a holon (there is no human body without its cells), there are, nonetheless, many instances where the upper-level holons of a system survive dramatic changes in the composition of a lower one. The smaller, faster, dynamics of elements nested at lower levels can therefore potentially drive innovations, innovations that can then, in the right circumstances, ramify throughout the system. On the other hand, the slower and vaster upper levels of the system can dampen, or absorb, such signals and because of this upper levels are more resistant to change. The more remotely separated are the levels of a system the less influence each one can exert upon the other. This tension of these dual tendencies – that of innovation and conservatism – in the language of systems theory this is sometimes labeled a system's capacity for *revolt* and *remembering* [40]. This is a concept that we return to towards the end of the book.

Putting some of the principles of the preceding paragraphs together with the use of a concrete example we might therefore say that on a landscape scale, (for example, at the scale of an entire tropical forest) the forested system may appear to be relatively stable, since the behavior of the forest is able to withstand dramatic changes that might be occurring on the plot scale. It is at this plot scale where the individual plant, for example, grows, is consumed, dies, germinates, and so on. Even though the behavior of the whole emerges from interactions among the parts, rapid and dramatic change at the smallest scale does not always influence the system in its entirety. The upper level (in this example, the landscape scale) imposes constraints on the environment experienced by elements contained within it. The climate, for example, may be stable at the level of the landscape, and this provides the relatively constant environment for individual organisms to complete their lifecycles. On the other hand, there are circumstances, especially when the change among the parts are synchronized, when constraints imposed from above are overwhelmed in a period of revolutionary change. In environmental systems the technical name for this hierarchical arrangement of levels – each of which have their own dynamics and their capacity to influence one another – is a *panarchy*, an important concept when considering stability and change in complex adaptive systems.

5.6 Emergent Properties

The patterns observed in any CAS are said to emerge from localized interactions occurring at lower levels in the system.

Taking a systems approach to the evaluation of environmental phenomena places an emphasis on the identification of the processes that determine the behavior of a system at the level defined by the observer [41]. The laws at the level of proximate interest to the investigator (*the focal level*) cannot violate the laws governing those that are lower in the same hierarchy [42]. Nevertheless, the phenomena manifest at the focal level may not be readily reduced to or understood as a summation of the properties of its parts.

Margaret Hilda Thatcher (1925–2013), former British Prime Minister, illustrates an extreme form of system reductionism in declaring: "And, you know, there is no such thing as society. There are individual men and women, and there are families." Thatcher, talking to *Women's Own* magazine, 31 October 1987. This statement transforms Baroness Thatcher into the most unlikely of systems thinkers. We might ask: from the perspective of hierarchy in what manner might this seem correct, and in what ways is she blatantly wrong. *Source:* Margaret Thatcher / Wikipedia Commons / CC BY-SA 3.0.

Ludwig Von Bertalanffy distinguishes two types of systems: *summative systems* and *constitutive systems*. A summative system can be adequately understood by examining the properties of its parts [43]. The functioning of a watch, for example, might be reasonably understood by assessing how its moving parts interact (though the smooth operation of watches can be quite complex) [42]. However, the properties of many systems are not so easily reduced. Even a system as seemingly simple as water with each molecule consisting of one oxygen and two hydrogen ions, has properties, such as its flow, and of being less dense as a solid than as a liquid that make it difficult to reduce the behavior of water to the properties of its parts. Most environmental systems are likewise considered to be "constitutive systems," where the dynamic interplay of the system elements contribute to behaviors that appear to be *greater than the sum of the parts*. The property of the whole being *greater than the sum of its parts*, is an essential element of a complex adaptive system [42].

Some of the features of the world that seem to most fascinate both scientists and nonscientists alike – consciousness, for example, and even the phenomenon of life itself – are emergent properties. The difficulty of satisfactorily explaining such properties by a reduction to the properties of the component parts, has often led some thinkers to speculate about mysterious forces – the soul, for example – to account for such phenomena. Arthur Koestler borrowing language from the philosopher Gilbert Ryle [1900–1976] refers to emergent properties, especially those of the mind, as the "ghost in the machine." This phrasing might seem to point to some mystical forces at play, but we should recognize that there is indeed a physical basis to these properties. They are nonlinear, and though not easily predictable, they are nonetheless investigable by scientific means [30].

The analysis of ecosystems, where the processes of biological production and decomposition are said to emerge from the functioning of complex mechanisms (operating in ecological communities that interact with the nonliving world), serves as an example of a research program based on constitutive properties that are ultimately amenable to scientific analysis.

5.7 An Important Aside of Determining System Scale in Science and Policy Collaboration

Determining the relevant system elements to include in any analysis of an environmental phenomenon, and the spatial and temporal scales on which that phenomenon should be inspected, is essential for annealing environmental research with the policy-making process. The fact that determinations of what to include and what to exclude (and where a system begins and ends) are difficult should come as no surprise. After all, ecological science's fundamental dictum – its first law – is that in an environmental system everything is connected. The great naturalist John Muir (1838–1914), phrases the dictum in this way: "when we try to pick out anything by itself, we find it hitched to everything else in the Universe" (see, *My First Summer in the Sierra* (1911) [44]). If everything is indeed connected then each environmental phenomenon may influence every other one. This being the case, what factors should one then include in any exploration of a system and, just as importantly, on what basis should some things be excluded? Including all the elements in a complex system may be impossible (given that we are finite beings), or at the very least, it becomes inefficient in practice since, as we have seen, some influences may be so negligible as to be irrelevant for system behavior. Because of this, most environmental models of CASs are necessary simplifications. That is, these models identify the most appropriate system elements and processes needed to adequately characterize a situation; as more data become available, the models characterizing it can become more refined.

To illustrate these challenges, consider investigations of *urban* ecological problems – and now, more than ever before, environmental disciplines are focused on urban settings. Urban phenomena will generally need to be framed in their appropriate social-ecological context, that is, both people (and institutional processes) and the processes of the physical and natural environment need to be considered in descriptions of such phenomena and in proposing solutions to very onerous municipal problems [45, 46].

Investigators need also be mindful of the temporal and spatial scales upon which a given problem is being addressed. The temporal and spatial scales of urban analysis should always include those that relate both to the sites of both production (e.g. food and resources) and consumption (i.e. where the food and resources are utilized). Thus, from an ecological perspective, cities rarely begin or end at their politically determined borders.

These examples from urban ecosystems can serve as an expression of a more general need to evaluate *all relevant elements* (and processes) in environmental systems analysis. Because of this, all participants in environmental analysis need high levels of proficiency in making determinations about the scope of such analyses. Considerable expertise is required in determining both system composition and its boundaries in practice but, at the very least, it is always important to report on the scales at which environmental analysis has been undertaken [47]. These decisions must often be determined subjectively; they certainly need careful deliberation.

5.8 Summary

In the above, we have provided an account of environmental phenomena as complex adaptive systems. The analysis of such systems involves identifying boundaries (both physical and dynamic), evaluating how constituent parts interact to produce complex wholes, and determining how the selective forces that act upon a CAS influences their development. Ultimately, humans are part of complex systems. How we humans are regulated as part of a complex adapted whole (that is, the global ecosystem), and yet can, in turn, impose our influence on the behavior of globe is the subject of much of what follows in subsequent chapters.

Throughout this chapter we have employed the term "environment," and you undoubtedly understood this work in the contexts in which it was used; nonetheless, the term is quite vague, and will need to be considerably sharpened before it can be operationized for use in research, or, indeed, in policy. This is the task of the next chapter.

References

1 Ashby, W.R. (1991). *Principles of the Self-Organizing System*, 521–536. Springer.

2 Karsenti, E. (2008). Self-organization in cell biology: a brief history. *Nature Reviews Molecular Cell Biology* 9 (3): 255–262.

3 Gell-Mann, M. (1995). *The Quark and the Jaguar: Adventures in the Simple and the Complex*. Macmillan.

4 Levin, S. (1999). *Fragile Dominion: Complexity and the Commons*. Reading, MA: Perseus.

5 Batten, D., Salthe, S., and Boschetti, F. (2008). Visions of evolution: self-organization proposes what natural selection disposes. *Biological Theory* 3 (1): 17–29.

6 Weber, B.H. and Depew, D.J. (1996). Natural selection and self-organization. *Biology and Philosophy* 11 (1): 33–65.

7 Holland, J.H. (2006). Studying complex adaptive systems. *Journal of Systems Science and Complexity* 19 (1): 1–8.

8 Harris, L.D. (1988). Edge effects and conservation of biotic diversity. *Conservation Biology* 2 (4): 330–332.

9 Rojas, J. et al. (2021). The micrometeorite flux at dome C (Antarctica), monitoring the accretion of extraterrestrial dust on earth. *Earth and Planetary Science Letters* 560: 116794.

10 Hurley, A.M. (2017). Dry stone wall building. https://www.teagasc.ie/rural-economy/rural-development/diversification/dry-stone-wall-building (accessed 16 November 2022).

11 Kranjc, A. (2009). Drystone Wall, an important element of karst cultural landscape: an example of dinaric karst. *The Egyptian Journal of Environmental Change* 1 (1): 8–12.

12 Robinson, T. (1986). *Stones of Aran: Pilgrimage*. Lilliput Press.

13 Haddon, A.C. and Browne, C.R. (1891). The ethnography of the Aran Islands, county Galway. *Proceedings of the Royal Irish Academy (1889–1901)* 2: 768–830.

14 Manenti, R. (2014). Dry stone walls favour biodiversity: a case-study from the Appennines. *Biodiversity and Conservation* 23 (8): 1879–1893.

15 Collier, M.J. (2013). Field boundary stone walls as exemplars of 'novel' ecosystems. *Landscape Research* 38 (1): 141–150.

16 Gibson, D.B. (2007). The hill-slope enclosures of the Burren, Co. Clare. *Proceedings of the Royal Irish Academy. Section C: Archaeology, Celtic Studies, History, Linguistics, Literature* 107C: 1–9.

17 Jones, C. (2020). Dry stone walls. In: *The Maintenance of Brick and Stone Masonry Structures* (ed. A.M. Sowden), 349–360. CRC Press.

18 Gould, S.J. and Lewontin, R.C. (1979). The spandrels of San Marco and the Panglossian paradigm: a critique of the adaptationist programme. *Proceedings of the Royal Society of London, Series B: Biological Sciences* 205 (1161): 581–598.

19 Darwin, C. (1871). Letter of Charles Darwin to J. D. Hooker, February 1871. www.darwinproject.ac.uk/letter/DCP-LETT-7471.xml (accessed 16 November 2022).

20 Mulkidjanian, A.Y. et al. (2012). Origin of first cells at terrestrial, anoxic geothermal fields. *Proceedings of the National Academy of Sciences* 109 (14): E821–E830.

21 Damer, B. (2016). A field trip to the Archaean in search of Darwin's warm little pond. *Life* 6 (2): 21.

22 Von Bertalanffy, L. (1967). General theory of systems: application to psychology. *Social Science Information* 6 (6): 125–136.

23 Gass, G.L. and Bolker, J.A. (2003). Modularity. In: *Keywords and Concepts in Evolutionary Developmental Biology* (ed. B.K. Hall and W.M. Olsen). Harvard University Press Credo Reference: https://ezproxy.depaul.edu/login?url=https://search.credoreference.com/content/entry/hupedb/modularity/0?institutionId=2536.

24 Levin, S.A. (1998). Ecosystems and the biosphere as complex adaptive systems. *Ecosystems* 1 (5): 431–436.

25 Gould, S.J. (1990). *Wonderful Life: The Burgess Shale and the Nature of History*. WW Norton & Company.

26 Horn, H.S. (1974). The ecology of secondary succession. *Annual Review of Ecology and Systematics* 5 (1): 25–37.

27 Pylyshyn, Z.W. (1998). Modularity of mind. In: *Routledge Encyclopedia of Philosophy* (ed. E. Craig). Routledge. https://doi.org/10.4324/9780415249126-W025-1.

28 Simon, H.A. (1991). *The Architecture of Complexity*, 457–476. Springer.

29 Koestler, A. (1970). Beyond atomism and holism — the concept of the Holon. *Perspectives in Biology and Medicine* 13 (2): 131–154.

30 Koestler, A. (1967). *The Ghost in the Machine*. London: Hutchinson.

31 Clune, J., Mouret, J.-B., and Lipson, H. (2013). The evolutionary origins of modularity. *Proceedings of the Royal Society B: Biological Sciences* 280 (1755): 20122863.

32 Novikoff, A.B. (1945). The concept of integrative levels and biology. *Science* 101 (2618): 209–215.

33 Allen, T.F. and Starr, T.B. (1982). *Hierarchy*. University of Chicago Press.

34 Holling, C.S. (1992). Cross-scale morphology, geometry, and dynamics of ecosystems. *Ecological Monographs* 62 (4): 447–502.

35 Feibleman, J.K. (1954). Theory of integrative levels. *The British Journal for the Philosophy of Science* 17: 59–66.

36 Turchin, P. (2001). Does population ecology have general laws? *Oikos* 94 (1): 17–26.

37 Begon, M., Harper, J.L., and Townsend, C.R. (1986). *Ecology. Individuals, Populations and Communities*. Blackwell Scientific Publications.

38 Tansley, A.G. (1935). The use and abuse of vegetational concepts and terms. *Ecology* 16 (3): 284–307.

39 Allen, T. and Hoekstra, T. (2015). *Toward a Unified Ecology*. Columbia University Press.

40 Abidi-Habib, M. and Lawrence, A. (2007). Revolt and remember: how the Shimshal nature trust develops and sustains social-ecological resilience in northern Pakistan. *Ecology and Society* 12 (2).

41 Levin, S.A. (1992). The problem of pattern and scale in ecology: the Robert H. MacArthur award lecture. *Ecology* 73 (6): 1943–1967.

42 Holland, J.H. (2014). *Complexity: A Very Short Introduction*. Oxford University Press.

43 Von Bertalanffy, L. (1969). *General System Theory: Foundations, Development, Applications*. George Braziller.

44 Muir, J. (1911). *My First Summer in the Sierra*. Houghton Mifflin.

45 Ostrom, E. (2009). A general framework for analyzing sustainability of social-ecological systems. *Science* 325 (5939): 419–422.

46 Ostrom, E. (2007). A diagnostic approach for going beyond panaceas. *Proceedings of the National Academy of Sciences of the United States of America* 104 (39): 15181–15187.

47 Nash, H.L. (2014). Defining appropriate spatial and temporal scales for ecological impact Analysis. *Environmental Practice* 16 (4): 281–286.

Section Three

The Concept of the Environment

6

All or Nothing? Or, What, Exactly, Is an Environment?

Though the concept of the "environment" has some antiquity, the word is used in neither Henry David Thoreau's *Walden* (1854) nor Charles Darwin's *On the Origin of Species* (1859), two texts regarded as foundational in the ecological sciences [1]. Certainly, by the time both of these works were written, the use of this word to denote the surroundings that affect the life of an organism was already established. However, the environment became a conceptual mainstay in natural history writing starting in the later nineteenth century, and it remains prevalent in contemporary ecological and nature-oriented literature [2]. In US-based opinion polls, concern about the "environment" consistently rank as being important [3], and it is generally in the sense of the external surroundings of organisms including humans that the word is being used. Though the word is readily understood in everyday use, nonetheless, conceptually, the "environment" has a vagueness about it, and the concept is in danger of becoming scientifically meaningless. In general, the problem with a definition of the environment that denote *all* the surrounding elements of an organism (or other biotic entity) is that it runs the risk of specifying *nothing* in particular.

A guinea pig in its natural environment (early eighteenth century). Note that in this picture the guinea pig is centered, and around the animal are arrayed a variety of visible resources (for instance, the vegetation), but also, less visibly, a range of conditions are implied: the air and its composition, the temperature, and so forth. The job of the ecologist is to discern those ecological factors most relevant to understanding the life history of the organism. Etching after J.B. Oudry. *Source:* Wellcome Images / Public Domain.

A Primer on Human Impacts on the Environment: The Conceptual Approach, First Edition. Liam Heneghan.
© 2023 John Wiley & Sons Ltd. Published 2023 by John Wiley & Sons Ltd.

Vague terms are not necessarily useless terms, as we have discussed in *Chapter Three*. In that chapter I alluded to the important philosophical literature devoted to such definitions; vagueness creates all sorts of interesting paradoxes: for example, how many grains of salt make a "heap," how much salt constitutes *more* than a "heap," and if you remove a single grain of salt does the heap cease to be a heap? (The philosophical problem of heaps, and otherwise blurry boundaries is referred to as the sorites paradox). In ordinary life, we muddle along despite the linguistic vagueness of many of our terms. However, for the purposes of doing both scientific and policy work there is a need for greater precision in the use of terms. Therefore, in this chapter I show how to operationalize the concept of the environment in quite specific ways. In particular, we will see how the concept is used in the basic ecological sciences when the factors shaping the lives of organisms are evaluated, and also how it can be employed to investigate an animal's *subjective* experience of the world, and, finally, how the human economy can be incorporated into our understanding of the total global environment. In these various ways we will have discovered how to navigate between the poles of *all* and *nothing*.

6.1   Vague Definitions of the Environment

The derivation of the term environment is from the Middle French word *environnement* meaning, "to surround something." Early use of the word signifies that it can, in addition, be employed to refer to the periphery, that is, to the side of something [4].

Though the environment is a key concept in the ecological sciences, many definitions of environment remain somewhat imprecise – and take very little account of the difficulties associated with implementation. A contemporary definition, given by the *Oxford English Dictionary*, and one that is close to how the word is used in the sciences is that the environment is "the physical surroundings or conditions in which a person or other organism lives, develops, etc., or in which a thing exists; the external conditions in general affecting the life, existence, or properties of an organism or object" [4]. The problem with this definition is as follows: if the environment is that which surrounds and influences an organism, then *everything* that is not the biological entity under immediate consideration can be regarded as potentially affecting the life and development of that organism. Even the definitions provided in standard dictionaries of biology and environmental management, where you might expect greater precision also suffer from the problem of being too broadly encompassing. For example, the environment is given by many such definitions as a "collective term for the conditions in which an organism lives, both biotic and abiotic," or as "the supporting matrices of life: water, earth, atmosphere and climate" [5, 6]. The inclusiveness of such definitions could indicate that, at best, environmental science entails a very extensive, possibly infinite, research program, but also a very diffuse one.

Compounding the problem associated with the vagueness of the concept of "environment" is the fact that one can also refer to both an "external environment" as well as an "internal environment." An internal environment refers to processes and their effects that occur within a body, though the term is more often explicitly evoked in psychology and physiology than in ecology [7]. In order to avoid stretching the word "environment" beyond reasonable limits, ecologists oftentimes refer to an "internal milieu" of a system or organism. It should, nonetheless, be clear that if we use the word environment to refer to everything both inside *and* outside an organism, the environment, despite the general heuristic appeal of the concept, is almost meaningless.

The concept of "environment" given by a vague definition can be made somewhat more explicit by restricting our attention to the "natural environment." Thus, the surroundings or conditions surrounding the organism or object are taken to be aspects of the physical and biological attributes of nature. However, the term "nature" itself, in turn, is one with a very broad meaning – that is, it is *also* a vague concept. The word "nature" can encompass all the phenomenon of the world in aggregate (including all entities and all natural processes): this is the meaning we use when we discuss chemistry, physics, and biology as "natural" sciences. Nature can also be used to indicate a particular place: that is, one can talk about going into the "world of nature," often in the sense of going to a particular often nonurban landscape or, more generally, off to the greener and wilder places of the world. Moreover, nature can signify the essence of something: when one strips away the superficial aspects of an object one supposedly gets to those unalterable and fixed qualities of the thing – that is, its true nature. Finally, nature may be used to designate all that is the opposite of culture. This distinction between *nature* and *culture* is a very commonplace meaning of the term, even though, from an evolutionary point of view, culture is a product of natural processes [8]. The list I have presented here on uses of the term nature is inspired largely by the work of historian Peter Coates whose useful book *Nature: Western Attitudes Since Ancient Times* (Berkeley, Univ. of California Press (1998) is recommended reading). The upshot of all of this is that referring to a "natural environment" compounds the problem of vagueness, since both nature and environment are all-encompassing, that is to say somewhat vague, terms.

The problem as we have set it out so far is this: both "environment" and "nature" are conceptually imprecise, and they can, therefore, be difficult to operationalize [9, 10]. What, then, should one study when studying the environment of a biological entity? We can sharpen our focus by recognizing that only certain selected aspects of the environment will be included in any investigation. These represent the *limited number of factors* that are most influential in shaping the life or development of an organism or community. The *limiting factors* that are of greatest interest to ecologists are the ones that constrain a population's size and can determine its growth.

Thus, the primary concern of environmental science (and of this book) is with how very particular aspects of the total environment shape the ecological and evolutionary success of living entities. As we have seen, ecological entities must be conceived of as operating across multiple scales. Living systems include individual organisms, which are part of a population of individuals of the same species, and as contributors to communities of multiple species, and, finally, these, in their turn are part of ecosystems. Thus the concept of the environment ultimately needs to be understood in terms of such systems existing on this variety of scales; factors that are relevant at one scale (for example, the microclimate influencing the life of a plant) may be less so on a larger scale. Our conceptualizing of the environmental will need to accommodate all of these scales.

A *biome* is the name given to major vegetation zones whose distribution is largely dictated by climatic factors (particularly temperature, precipitation, and, consequently, soil moisture). Each region supports plants with a set of characteristic life forms. Some of the major biomes include tundra, boreal forest, temperate forest, temperate grasslands, Mediterranean-type vegetation desert, tropical grasslands, tropical seasonal forest, and tropical rain forest. What is important to notice is that the distribution of biomes is controlled not by the total environment, but by a relatively small number of environmental factors.

Ecologists who want to understand the distribution of biological communities on large geographical scales will typically not evaluate each and every ecological factor of the environment, but rather will initially evaluate temperature, rainfall, and soil properties, as those aspects of the environment most likely to influence large-scale patterns of distribution. In this way, the ecologist can predict the geographical occurrence of tropical forest, tallgrass prairie, tundra, and so on with relatively minimal knowledge of environmental conditions. This grouping of plant communities according to their shared adaptations to environmental factors is referred to as a *biome* – that is, the regional ecological complex that is often climatically controlled – a concept that has stood the test of time in ecology [11, 12].

In order to understanding the distribution of organisms on very fine scales it is necessary for the ecologist to know the full panoply of potentially influencing factors that limit both the distribution and fitness of organisms. Depending upon the specific goals of a project, the scientist will draw from this palette of factors in designing a study. Conceptually this range of options can be referred to as the "environmental mandala," which we will now consider. The mandala is one very powerful way to operationalize the notion of an environment.

6.2   The Environmental Mandala

To operationalize the traditional notion of the environment for use in ecological investigations, imagine the following: an organism or collection of organisms (a population or community, for example) is figuratively placed at the center of a circle around which are arrayed all of the elements of the encircling world. Of course, myriad factors can directly affect the life and development of the organism over the course of that entity's development. Being finite, an ecologist can rarely consider all of them, but, as we have seen, not all these potentially explanatory variables are necessary in answering most ecological questions. A narrow range of such factors, for example, key nutrients, access to sunlight, and availability of moisture, are most determinative for the life of the organism. A range of biotic factors: predators, competitors, and symbionts needs also to be investigated. One can think of this process of conceptually whittling down the entire world to a list of the more determinative factors as being like "imagining a spherical cow," which in an earlier chapter we noted provides a guiding framework for model building. This whittling process simplifies model construction and makes investigating an ecological phenomenon more tractable. Indeed, such simplification is the essence of conceptual modeling.

The late American ecologist Frank Golley (1930–2006) called the diagram of all possible influences on the survival of an organism, an "environmental mandala," naming it after Buddhist and Hindu geometric symbols of the cosmos [10]. While the explicit terminology of an "environmental mandala" has not been especially influential, nonetheless, a mandala reminds us that ecology addressed three essential operational tasks:

1) Determining what entity to place at the *center of mandala*. That is, what are the appropriate units of environmental analysis?
2) Evaluating the *relative importance of an array factors* in the roster of all biotic and abiotic ones that potentially determine the survival and thriving of the chosen unit?
3) Establishing *the reciprocal influence of entity and environment*: that is, we can recognize the environmental factors that determine the vigor of the entity centered in the mandala, but need also to evaluate how the living entity, in turn, can change the availability of those factors for other organisms. Environmental factors made less available after being appropriated by individual organisms are referred to as "ecological resources". A change in resources will alter the mandala and affect neighboring organisms. The mandala is therefore dynamic.

It is easiest to conceive of the entity at the center of the mandala as the *individual organism*. In ecology, the investigation of single organisms is referred to as *autecology*. More often than not, an individual organism inspected by a scientist is an exemplar drawn from a *population of organisms* (that is, the individual organism is rarely inherently interesting to the ecologist). A population is defined as a system comprised of the individuals of a single species found in a location at one particular time. As both biotic and abiotic environmental factors vary over time, the abundance of a population will increase or decrease [13]. An environmental scientist tries to determine where organisms are most likely to flourish within that species' geographical range [14]. Explaining the distribution of organisms at grosser landscape scales will require an evaluation of a relatively small number of environmental factors. As the scale of inspection becomes more refined – for example, to explain why an individual organism is found in one very particular location – the number of factors needed to address these *fine-grained patterns* in the distribution can become very large indeed [15].

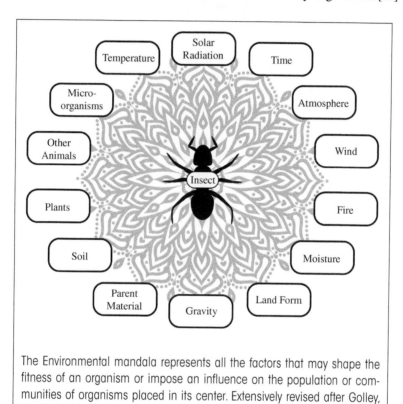

The Environmental mandala represents all the factors that may shape the fitness of an organism or impose an influence on the population or communities of organisms placed in its center. Extensively revised after Golley, F.B. (1998), by Sarah Heneghan.

A system comprised of individual organisms from several species found simultaneously in the same place is referred to as a *biotic community*. In terms of the mandala, one can conceive the community as constituting several elements of the mandala (animals, plants, microbes) that interact and impose a mutual influence. Any organism in the mandala can occupy its center, but the role of the organism in reference to other organisms in the community often becomes the key question. The environmental regulation of communities of organisms has been a central question in the environmental sciences for over a century and a truly dizzying number of concepts seeking to classify communities of organisms have been proposed [16]. A spirited debate occurred in the decades of the last century about the "ontological status" of these communities, that is, wondering whether plant communities, in particular, should be considered "real" in the way that, say, an individual plant is considered "real" [17]. Interesting though such debates are, the issue for us is to recognize here is that once they are aggregated into communities, multispecies assemblages then become the subject of environmental analysis. How do both biotic and abiotic environmental factors influence the composition and distribution of such communities? As we have seen from our brief consideration of the biome model above, the number of factors that need to be considered may be, in some circumstances, relatively few, and those are mostly related to regional climate and to both soil composition and the geological substrate upon which the community is established. The concept of the biome is most often applied to plants but can be used in understanding patterns of animal communities [18].

One final ecological unit relevant to operationalizing the concept of environment, is that of the *ecosystem*, introduced by Arthur Tansley in 1935 [19]. Tansley introduces this concept in the following manner:

> But the more fundamental conception is, as it seems to me, the whole system (in the sense of physics), including not only the organism-complex, but also the whole complex of physical factors forming what we call the environment of the biome—the habitat factors in the widest sense.

From the outset, the appeal of the ecosystem concept is that it considers all the organisms in a community alongside those physical (abiotic) factors that are consequential for survival. In an important sense the ecosystem is just another name for the entire environmental mandala, especially when a study focuses on key processes such as organic matter decomposition or nutrient cycling. The concept of the ecosystem unsurprisingly has itself been considered vague and difficult to operationalize [20]. Indeed, Tansley seems to have insisted on the conceptual nebulousness of the ecosystem when he wrote, "These ecosystems, as we may call them, are of the most various kinds and sizes. They form one category of the multitudinous physical systems of the universe, which range from the universe as a whole down to the atom" [19].

To avoid problems of vagueness, ecosystem ecologists have operationalized the ecosystem concept in a number of ways. Firstly, the ecosystem is often examined as a spatial unit. Such units can be, for example, a cavity in a tree, a log, a lake, or an individual stand of forest [21, 22]. Conceiving of ecosystems as watersheds – that is, the entire drainage basin of a stream, river, or lake – had been especially helpful since this facilitates studies of systems that are, to some extent, isolated from adjacent systems. Furthermore, researchers can install instruments in discrete watersheds to measure water flow, dissolved nutrients, sediments, and other outputs from the terrestrial component of these landscape units [23].

The move to make the ecosystem concept more concrete by spatializing it has fostered an interest in exploring the properties of artificially constructed and simplified ecosystems in the laboratory or in the field. If an ecosystem can be a log, surely it can be a small microcosm constructed in the laboratory, or a closed ecological system. An example of an artificial system used in research is the "ecotron" – a series of sixteen controlled environmental chambers at the Natural Environment Research Council (NERC) Centre for Population Biology in the United Kingdom [24]. Despite, the enthusiasm for this approach to researching ecosystems, there has been persistent criticism that such systems are too simplistic to be really useful [25].

6.3 The Umwelt: The "self-world", as Environment

The environmental mandala conceptually centers a biotic entity (organism, population community, or ecosystem) in an environing circle and investigates those factors, both biotic and abiotic, that regulate the life of that entity. Of course, it is easier to conceive of the mandala with an individual organism as it focus. As one expands the scale from community to ecosystem, the ecologist (or manager) focuses upon several biotic and abiotic elements simultaneously. The strength of the approach ultimately is that the mandala makes the vague concept of environment something more concrete that can then be investigated systematically. The approach is objective in the sense that it seeks generalizable statements about the environmental influence on the focal biotic entity, and, reciprocally, on how that entity changes its environment.

Another approach to operationalizing the idea of the environment comes from investigating the *subjective* experience that an individual animal has of their environmental circumstances. Estonian biologist Jakob von Uexküll (1864–1944) proposed the concept of an *Umwelt* (a German word that can be translated as "self-world") to describe the perceptual world (or "bubble") of an organism [26]. In this view of things, each organism occupies and responds to its own perceived environment. Each animal is in its own "self-world" and uses the sense receptors characteristic of its species to both encounter and make meaning of its world. Information about an object encountered by the animal in its environment is processed in the Umwelt, and, once processed, a signal is conveyed to effector organs which can discharge an appropriate response. Accordingly, by means of its sensory and effector organs each creature inhabits a portion of its objective (physical–chemical–biological) surroundings. Unlike definitions of the environment that comprise *everything* in the universe surrounding the organism, the Umwelt model includes *nothing*, until the organism lights up that thing with meaning by processing and responding to the object in its world.

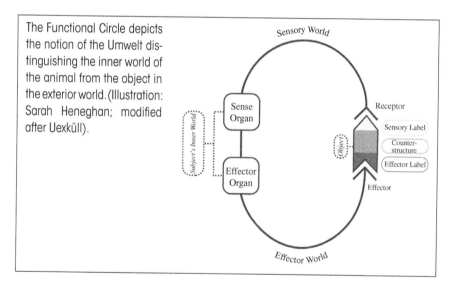

The Functional Circle depicts the notion of the Umwelt distinguishing the inner world of the animal from the object in the exterior world. (Illustration: Sarah Heneghan; modified after Uexküll).

A depiction of the connection between an object in the environment of an animal that has been detected by means of its sensory receptors and responded to by its effector organs is referred to as a *functional circle*. An object in the world does not become part of the environment until it is engaged by the functional circle. Like the mandala, the Umwelt of an organism can be depicted as occupying a fixed sphere. However, in the case of the Umwelt, that sphere is created by the individual as it lives its life. That world can be discovered by researchers through close observation of the animal's behavior. Using knowledge of the functional cycle – that is, that sequence of steps that connects objects to the perceptual world and responsive actions appropriate to that organism – a scientist can gain access to the Umwelt of the organism. For this reason, Uexküll's concept of Umwelt has been particularly influential in animal behavior studies (ethology), psychology, and in certain elements of twentieth century philosophy. The Umwelt is also a valuable concept in the field of biosemiotics, that is, the study of how signs and codes in the biological world are translated in meaningful information. It should be regarded as a complementary approach to that of the mandala, not an alternative.

Although the umwelt has been applied to a variety of organisms (Uexküll's examples range from ticks, planarian worms, sea urchins, and all the way up to the higher animals), it may be especially useful in thinking about the environment of humans. Humans utilize aspects of their environing world in extremely various ways, and, furthermore, can confer a range of different meanings upon the same object: a daffodil can be regarded as ecologically meaningful, but can also serve, more poetically, as a harbinger of spring. Though all able-bodied humans roughly share the same sensory and effector equipment, each individual occupies to some extent a separate Umwelt. The individuated Umwelt is defined by the objects each of us encounter and the ways in which we respond to these interactions. Each person has a distinctive perceptual and operational world, but, additionally, we share a social world of shared meaning. One of the consequences of living in this present moment of time where human occupy most parts of the planet, is that innumerable objects in the global environment have now become part of the Umwelt of the human. Much of the outer-world, therefore, has now become part of the inner-world of the human being: we have, for better or worse, spread a net of meaning across the entire globe.

6.4 Human/Environment Interactions: Standard Models

Human environmental interactions can be depicted using the same conceptual tools that are applied to any other organism. When we apply a natural history approach to humans, we place ourselves at the center of the mandala, and by doing so, we can assess how human flourishing depends upon diverse environmental factors. Furthermore, by simultaneously taking up so many parts of the world into the human Umwelt, we confer a wide range of meanings to several aspects of the global environment. A variety of systems models have been developed to try to make sense of human relationships with the rest of the natural world. In the crudest version of such models, researchers isolate the human component and assess our collective dependence upon the rest of nature. Thus, a human realm can be distinguished from that of "natural" ecosystems. The quotation marks around the word "natural" indicates that though humans can be extracted from nature for the purposes of analysis, humans are indissolubly part of the natural world. It can nevertheless be analytically useful for a model to distinguish *human dominated* from *less-dominated* or *"natural" systems* [27, 28]. Such models as these then consist of three compartments: a human realm that is distinguished from a "natural" (or less-dominated) one and the global ecosystem in which the first two components are embedded. Despite the simplicity of this representation, several important observations can be made about such models.

For practical purposes global ecosystems can be considered *materially closed*, but *energetically open*. Although our planet, as we have already mentioned, receives a constant input of matter in the form of cosmic dust, meteorites, and so on, the mass of this material is very small in comparison with the mass of the planet [29]. The material resources of the planet can therefore essentially be regarded as finite. Because of this, the primary elements needed for living beings must to be recycled – and humans are dependent upon this recycling of key nutrients while also contributing to and amplifying naturally occurring cycles of these elements. The biogeochemistry of carbon, hydrogen, nitrogen, oxygen, phosphorus, and sulfur are of great significance in ecology. These elements are important alongside a trailing edge of other elements that are needed by living beings in small quantities (such as sodium, magnesium, potassium, calcium, and chlorine). The human influence on the cycling of several of these elements – carbon, nitrogen and phosphorus, in particular – are intensively studied because their modification by human activity is significant for planetary functioning. Nitrogen is often described as the "limiting" factor for terrestrial productivity, and phosphorus plays a comparable role in aquatic systems (though both of these elements can strongly influence growth in both terrestrial and aquatic systems).

> ***Ecological production*** is typically measured as the amount of carbon biomass (that is, stored chemical energy) accumulated by photosynthesizing organisms (the primary producers) over a given period. The total production is referred to as *Gross Primary Productivity* (GPP). The fraction that remains when the cost of respiration and the maintenance of existing tissues of the primary producers is subtracted is referred to as *Net Primary Production* (NPP). Though production ultimately depends upon sunlight as an energy source, the availability of nutrients (including nitrogen and phosphorus) is crucial.

All of the ecological processes that organisms engage in incur energetic costs. The study of ecological energetics is therefore a necessary (though sometimes neglected) subdiscipline, one that is required for understanding humans in a changing world [30]. Certain bacteria can synthesize organic molecules by means of the oxidation of specific inorganic molecules or ions. Such bacteria are often joined in chemosynthetic symbioses with marine invertebrates. These symbiotic pairs are found in a range of habitats, for example from the deep sea, cold seeps, shallow-water coastal sediments and continental margins [31]. Though chemosynthesis may have been crucial to the origins of life and the early evolution of the biosphere, photosynthesis prevails in contemporary ecosystems [32, 33]. The flow of energy by means of photosynthetic fixation, animates the exchange of key nutrients. The crucial thing to note here is that the Earth is open to the energy of the sun, and this is the energy that drives the cycles of finite material upon which life relies.

When constructing these models it is also useful to take account of the spatial extent of the human economy. Such analysis should encompass all of the lands that lie in the shadow of human use. (This spatial unit is sometimes labeled the "ecological footprint," as will be discussed in a later chapter). A suite of important processes connect the human dominated spatial components of landscapes with residual "natural" components (or the lands less affected by humans). These connections result in a flow of goods and other provisions termed "ecosystem services." Ecosystem services include the multiple benefits provided by ecosystems to humans – pollination, nutrient cycling, soil formation, the provisioning of fresh air, and many more [34]. Such services, according to the analytical framework used in *The Millennium Ecosystem Assessment Synthesis Report* [34] include provisioning services, regulating services, cultural services, and the supporting services, upon which the first three categories depend. As the human population grows and the consumption of resources increases, the provisioning of ecosystem services may, in some circumstances, decrease. Declining soil fertility, altered water availability, diminished mineral supply, and so on, in turn, can have impacts upon human flourishing [35]. The vulnerability of ecosystem services to human impacts is not always a direct one – that is, the erosion of such services occurs not only when lands are under immediate human management or when these lands have been converted from one purpose to another. In addition to direct influences, the provisioning of ecosystem services may also be diminished by an array of indirect factors, for example, those emanating from climate change or from the loss of wildlife [36].

6.5 Modifying the Basic Model: Social Ecological Models

The models that reveal connections between humans and the rest of nature recognize that, despite our very distinctive attributes, needs, and the particular wastes that we produce, humans, like all other organisms, are ultimately embedded within the global ecosystem. This is true despite our living, for the most part, in habitats constructed for the immediate needs of people (including our dwellings, as well as our towns and cities). In the most basic of these models, all of the forces that shape human life and well-being – including both our physical dependence upon nature and our social dependence – are lumped together. The human individual is molded as much by our social attributes – that is, by an array of cultural, economic, and familial structures that determine the content of all of our lives – as it is by our ecological situation. Thus, human lives can be conceived as being in the grip of institutional forces, and as a result the processes by which social decisions are made must also be considered in environmental analysis. Acknowledging that humans require the tangible resources provided by ecosystems, while also being dependent on social structures and practices has resulted in the development of explicitly *social ecological models* [37, 38]. Exemplary models of this kind, which depict the influence of both social and environmental factors on the health and well-being of the individual, had initially the most striking impact on the field of child development and early education. The developmental theories of Urie Bronfenbrenner, a Russian-born American psychologist, systematically presented several such models [39]. More recently, environmental policy has been influenced by the research of Carl Folke (b. 1955), a Swedish interdisciplinary researcher with training in economics and ecology. Folke, together with his colleagues both in Sweden and internationally, uses social-ecological models to illustrate how processes such as "social learning and social memory, mental models and knowledge" and other social-institutional attributes need to be integrated alongside overtly ecological factors in the management of essential ecosystem services [40].

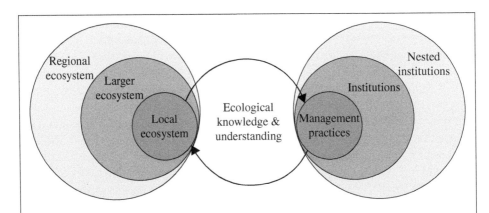

Social-ecological systems analysis combines powerful tools from the social sciences that examine governance and management practices with those from the ecological sciences evaluating the patterns of nature and its provisioning of goods and services to the human economy. The figure is adopted from Colding, Johan, and Stephan Barthel. "Exploring the social-ecological systems discourse 20 years later." Ecology and Society 24, no. 1 (2019). Open Access statement: https://ecologyandsociety.org/policies/.

There are myriad ways of depicting social-ecological systems (SES) and all have the following in common: they integrate both social and ecological components (and their associated dynamics) into a unified framework. However, the depth to which both the social and the ecological are treated in these models can vary. Binder and colleagues have distinguished ten styles of these models [41]. For completeness I will list these models (and I recommend the reader to follow up on each to understand their specific uses): *Driver, Pressure, State, Impact, Response* (DPSIR), *Earth System Analysis* (ESA), *Ecosystem Services* (ES), *Human Environment Systems Framework* (HES), *Material and Energy Flow Analysis* (MEFA), *Management and Transition Framework* (MTF), *Social Ecological System Framework* (SESF), *Sustainable Livelihood Approach* (SFA), *The Natural Step* (TNS), and *Turner's Vulnerability Framework* (TVF) models. Each of these models is designed for a particular purpose. For example, the SESF was developed be Elinor Ostrom, a Nobel laureate economist, and her colleagues, and it provides a framework for analyzing the relationship between governance systems and extractive resources [42]. The Natural Step (TNS) framework, complementarily, provides tools for sustainability analysis.

Elinor (Lin) Ostrom (1933–2012) a leading American political economist whose work on social ecological modeling continues to be very influential in social ecological modeling. In 2009, Ostrom was awarded the Nobel Memorial Prize in Economic Sciences, the first woman to win that prize. Holger Motzkau / Wikimedia Commons / CC BY-SA 3.0.

In general, social-ecological models are especially influential in the emerging subdiscipline of urban ecology [43–45]. Ecologists Steward Pickett and J. Morgan Grove illustrate the elements involved in an integrated social ecological model which is applied to urban situations [46]. In their model, land use, land cover, production, consumption, and the disposal and flux of wastes are viewed as emerging from interactions between social and bio-ecological components of the system. The way in which the patterns and processes associated with a given urban landscape emerge must be understood in the context of the ecological, political, and economic circumstances governing the external conditions of the city.

Models depicting the Anthropocene – the name proposed for the contemporary geological era in which the activity of humans exert unprecedented impacts on a global scale – are inherently social and ecological simultaneously. These models are the subject of a later chapter in this book.

6.6 Modifying the Basic Model: Planetary Sources and Sinks

Social-ecological models can be adapted for a variety of purposes. For example, one can use it to conceptualize differences in patterns of resource use by indigenous societies compared with that of contemporary post-industrial societies or we can examine changes in any one society over time. In an important adaptation of the model, ecological economist Herman Daly (echoing, as we shall see, modeler Donella Meadows and her team) recognizes that the human economic component of global systems draws upon resource stocks and produces wastes as a result of transformation by economic activity [47]. The distinction between planetary sources and planetary sinks is terminology we will employ extensively in future chapters. In this way, ecological economists distinguish the flows of both material and stored energy needed to sustain the economy, from the flows of waste and pollution that issues from the economy and that end up in a variety of planetary sinks (water, soils, and the atmosphere, in particular).

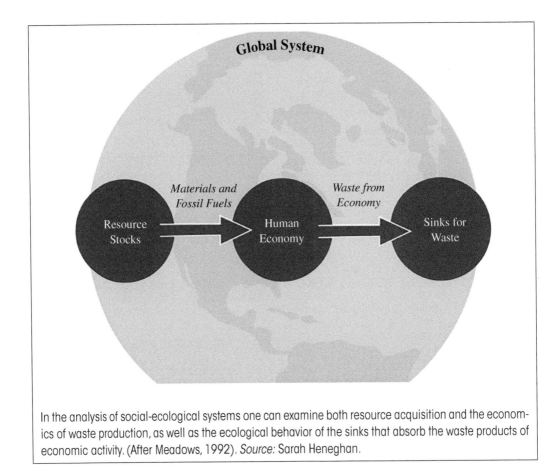

In the analysis of social-ecological systems one can examine both resource acquisition and the economics of waste production, as well as the ecological behavior of the sinks that absorb the waste products of economic activity. (After Meadows, 1992). *Source:* Sarah Heneghan.

Thought of in this way, one can evaluate the sources of environmental problems confronting the global economy as deriving from *either* resource shortage or from the befouling of ecological sinks. Examples of the former type of problem include many of the traditional issues addressed by environmental policy makers: concern for the conservation of timber [48], general concerns about a dwindling supply of natural resources including availability of freshwater [49, 50], exhaustion of fossil fuels [51, 52], and estimates of the extent of human appropriation of primary productivity (that is, what percentage of energy captured by plants is diverted to the needs of just one species: humans) [53, 54].

Problems with sinks, on the other hand, can be illustrated by considering the following environmental problem. A body of water – for example, a lake – will typically have the capacity to absorb some nutrient pollution, emanating, for instance, from excessive phosphorus run-off.

This phosphorus loading can come from surrounding farmlands or from a heavily fertilized domestic landscape. Up to a limit, the lake can absorb these wastes without a significant impairment in its functioning. However, the capacity of a sink such as a lake can be exhausted after which additional input can overwhelm the system, resulting in dramatic shifts in the structure and functioning of the system. The exhaustion of sinks, considered at the level of the globe can result in a transformation of planetary function. This large claim is one that we investigate in later chapters. Thus, for example, the atmosphere acts as a global sink for carbon dioxide, and up to a certain point the behavior of the atmosphere (reflected in climate) is unaffected by modest increases in inputs from human activities such as the burning of fossil fuels [55]. Problems associated with significant levels of climate change can be conceived as expressing the exhaustion of a global sink. Considering this claim will be the subject of Chapter 12.

6.7 Full World Versus Empty World

The innovation of ecological economics, a relatively newly-founded discipline, is that it embeds the structures of the human economy into the rest of nature [56, 57]. Therefore, unlike in neoclassical economics, which focuses on the behavior of rational actors functioning within markets, ecological economics depicts the behavior of consumers and firms as being constrained within the limits of a global environment. Ultimately, ecological economics examines the possibility of a global transition away from ecologically unwise practices to a steady state or sustainable economy [58–61]. Such an economy would provide for the flourishing of humans without compromising the capacity of the global environment to provide for future generations, and would prioritize doing so in an equitable way.

When the scale of human economic activity is evaluated in the context of global limits, an ecological economist can query the degree to which human activity has reached the limits of the planet. Using language that Herman Daly developed in *Beyond Growth: The Economics of Sustainable Development* (1996) we can distinguish an "empty world" scenario, where the scale of economic activity is small compared to the capacity of the global ecosystem, from a "full world" scenario where humans are bumping up against the limits of the planet.

Whether or not we are occupying a full world or an empty one is the topic of central concern in subsequent chapters.

6.8 Summary

In this chapter, we started with the problem of taking the somewhat vague concept of the environment and operationalizing it. This can be done in a variety of ways, though the essence of *all* models that operationalize the environment is their necessary simplification. The environmental researcher must identify the limited number of variables that can be most influential for individual organisms, the populations they comprise, the communities in which they find themselves, and the ecosystems they constitute. These factors must be distinguished from among a vast range of ecological factors, many of which, admittedly, will be trivial in their influence. The sort of models developed for studying the environment of non-human organisms can be adapted and applied to humans. In models of this sort the fact that humans have both physical needs (typically met by the natural environment) and social needs (met as part of a community of our fellow humans) is recognized, and incorporated into social-ecological frameworks.

References

1 Worster, D. (1994). *Nature's Economy: A History of Ecological Ideas*. Cambridge University Press.

2 Graham, M.H., Parker, J., and Dayton, P.K. (2011). *The Essential Naturalist: Timeless Readings in Natural History*. University of Chicago Press.

3 Environment. Gallup. https://news.gallup.com/poll/1615/environment.aspx (accessed 16 November 2022).

4 environment, n. OED Online. (2022). Oxford University Press. https://www-oed-com.ezproxy.depaul.edu/view/Entry/63089?redirectedFrom=environment (accessed 27 June 2020).

5 Thain, M. and Hickman, M. (2004). *Penguin Dictionary of Biology*. Penguin Books.

6 Calow, P.P. (ed.) (1998). *The Encyclopedia of Ecology and Environmental Management*. Blackwell Science. Blackwell Publishers environment.

7 Reber, A.S., Reber, E., and Allen, R. (2009). *The Penguin Dictionary of Psychology*. Penguin *internal environment*. Penguin Books.

8 Coates, P. (2013). *Nature: Western Attitudes Since Ancient Times*. Wiley.

9 Morton, T. (2009). *Ecology without Nature: Rethinking Environmental Aesthetics*. Cambridge, MA: Harvard University Press.

10 Golley, F.B. (1998). *A Primer for Environmental Literacy*. New Haven, CT; London: Yale University Press.

11 Carpenter, J.R. (1939). The biome. *American Midland Naturalist* 21 (1): 75–91.

12 Thain, M. and Hickman, M. *Penguin Dictionary of Biology*, 11e. London: Penguin biome.

13 Sibly, R.M. and Hone, J. (2002). Population growth rate and its determinants: an overview. *Philosophical Transactions of the Royal Society of London. Series B, Biological Sciences* 357 (1425): 1153–1170.

14 Turkington, R. and Harper, J.L. (1979). The growth, distribution and neighbour relationships of *Trifolium repens* in a permanent pasture: IV. Fine-scale biotic differentiation. *The Journal of Ecology* 245–254.

15 Rietkerk, M., Ouedraogo, T., Kumar, L. et al. (2002). Fine-scale spatial distribution of plants and resources on a sandy soil in the Sahel. *Plant and Soil* 239 (1): 69–77.

16 Whittaker, R.H. (1962). Classification of natural communities. *The Botanical Review* 28 (1): 1–239.

17 Loehle, C. (1988). Philosophical tools: potential contributions to ecology. *Oikos* 51: 97–104.

18 Odum, E.P. (1945). The concept of the biome as applied to the distribution of North American birds. *The Wilson Bulletin* 57 (3): 191–201.

19 Tansley, A.G. (1935). The use and abuse of vegetational concepts and terms. *Ecology* 16 (3): 284–307.

20 Golley, F.B. (1993). *A History of the Ecosystem Concept in Ecology: More than the Sum of the Parts*. Yale University Press.

21 Lindeman, R.L. (1942). The trophic-dynamic aspect of ecology. *Ecology* 23 (4): 399–417.

22 Walker, E.D., Lawson, D.L., Merritt, R.W. et al. (1991). Nutrient dynamics, bacterial-populations, and mosquito productivity in tree hole ecosystems and microcosms. *Ecology* 72 (5): 1529–1246.

23 Likens, G.E. (2013). *Biogeochemistry of a Forested Ecosystem*. Springer Science & Business Media.

24 Lawton, J.H. (1996). The Ecotron facility at Silwood Park: the value of "big bottle" experiments. *Ecology* 77 (3): 665–669.

25 Carpenter, S.R. (1996). Microcosm experiments have limited relevance for community and ecosystem ecology. *Ecology* 77 (3): 677–680.

26 Uexküll (1921). *Umwelt und innenwelt der tier*. Berlin, Heidelberg: Springer.

27 Carter, N.H., Vina, A., Hull, V. et al. (2014). Coupled human and natural systems approach to wildlife research and conservation. *Ecology and Society* 19 (3): 43.

28 Liu, J.G., Dietz, T., Carpenter, S.R. et al. (2007). Complexity of coupled human and natural systems. *Science* 317 (5844): 1513–1516.

29 Love, S.G. and Brownlee, D.E. (1993). A direct measurement of the terrestrial mass accretion rate of cosmic dust. *Science* 262 (5133): 550–553.

30 Tomlinson, S., Arnall, S.G., Munn, A. et al. (2014). Applications and implications of ecological energetics. *Trends in Ecology & Evolution* 29 (5): 280–290.

31 Dubilier, N., Bergin, C., and Lott, C. (2008). Symbiotic diversity in marine animals: the art of harnessing chemosynthesis. *Nature Reviews. Microbiology* 6 (10): 725–740.

32 Campbell, K.A. (2006). Hydrocarbon seep and hydrothermal vent paleoenvironments and paleontology: past developments and future research directions. *Palaeogeography, Palaeoclimatology, Palaeoecology* 232 (2–4): 362–407.

33 Ryu, Y., Berry, J.A., and Baldocchi, D.D. (2019). What is global photosynthesis? History, uncertainties and opportunities. *Remote Sensing of Environment* 223: 95–114.

34 Leemans, R. and De Groot, R. (2003). *Millennium Ecosystem Assessment: Ecosystems and Human Well-Being: A Framework for Assessment*. Island Press.

35 Schroter, D., Cramer, W., Leemans, R. et al. (2005). Ecosystem service supply and vulnerability to global change in Europe. *Science* 310 (5752): 1333–1337.

36 Weiskopf, S.R., Rubenstein, M.A., Crozier, L.G. et al. (2020). Climate change effects on biodiversity, ecosystems, ecosystem services, and natural resource management in the United States. *The Science of the Total Environment* 733: 137782.

37 Holling, C.S. (2001). Understanding the complexity of economic, ecological, and social systems. *Ecosystems* 4 (5): 390–405.

38 Ostrom, E. (2009). A general framework for analyzing sustainability of social-ecological systems. *Science* 325 (5939): 419–422.

39 Bronfenbrenner, U. (1992). *Ecological Systems Theory*. Jessica Kingsley Publishers.

40 Folke, C. (2006). Resilience: the emergence of a perspective for social-ecological systems analyses. *Global Environmental Change-Human and Policy Dimensions* 16 (3): 253–267.

41 Binder, C.R., Hinkel, J., Bots, P.W.G., and Pahl-Wostl, C. (2013). Comparison of frameworks for analyzing social-ecological systems. *Ecology and Society* 18 (4): 26.

42 Ostrom, E. (2011). Background on the institutional analysis and development framework. *Policy Studies Journal* 39 (1): 7–27.

43 Collins, J.P., Kinzig, A., Grimm, N.B. et al. (2000). A new urban ecology. *American Scientist* 88 (5): 416–425.

44 McPhearson, T., Pickett, S.T.A., Grimm, N.B. et al. (2016). Advancing urban ecology toward a science of cities. *Bioscience* 66 (3): 198–212.

45 Wu, J.G. (2014). Urban ecology and sustainability: the state-of-the-science and future directions. *Landscape and Urban Planning* 125: 209–221.

46 Pickett, S.T. and Grove, J. (2009). Urban ecosystems: what would Tansley do? *Urban Ecosystems* 12 (1): 1–8.

47 Daly, H.E. (1996). *Beyond Growth: The Economics of Sustainable Development*. Beacon Press.

48 Pinchot, G. (1910). *The Fight for Conservation*. Doubleday, Page.

49 Pimentel, D., Harman, R., Pacenza, M. et al. (1994). Natural resources and an optimum human population. *Population and Environment* 15 (5): 347–369.

50 Postel, S.L., Daily, G.C., and Ehrlich, P.R. (1996). Human appropriation of renewable fresh water. *Science* 271 (5250): 785–788.

51 Kunstler, J.H. (2007). *The Long Emergency: Surviving the End of Oil, Climate Change, and Other Converging Catastrophes of the Twenty-First Century*. Open Road+ Grove/Atlantic.

52 Klare, M. (2002). *Resource Wars: The New Landscape of Global Conflict*. Macmillan.

53 Haberl, H., Erb, K.H., Krausmann, F. et al. (2007). Quantifying and mapping the human appropriation of net primary production in earth's terrestrial ecosystems. *Proceedings of the National Academy of Sciences* 104 (31): 12942–12947.

54 Vitousek, P.M., Ehrlich, P.R., Ehrlich, A.H., and Matson, P.A. (1986). Human appropriation of the products of photosynthesis. *BioScience* 36 (6): 368–373.

55 Pacala, S.W., Hurtt, G.C., Baker, D. et al. (2001). Consistent land-and atmosphere-based US carbon sink estimates. *Science* 292 (5525): 2316–2320.

56 Costanza, R., Cumberland, J.H., and Daly, H. (ed. et al.) (2014). *An Introduction to Ecological Economics*. CRC Press.

57 Daly, H.E. and Farley, J. (2011). *Ecological Economics: Principles and Applications*. Island Press.

58 Daly, H.E. (1973). *Toward a Steady-State Economy*. San Francisco: W.H. Freeman.

59 O'Neill, D.W. (2012). Measuring progress in the degrowth transition to a steady state economy. *Ecological Economics* 84: 221–231.

60 Munasinghe, M. (1993). *Environmental Economics and Sustainable Development*. The World Bank.

61 Pearce, D.W. and Warford, J.J. (1993). *World without End: Economics, Environment, and Sustainable Development*. Oxford University Press.

7

Life and Environment Are Indissolubly Linked

Supposing you stumbled upon an ancient rock bearing fossils of these creatures – trilobites (an extinct group of early marine arthropods) – how would you verify that they were formerly alive? You might reflect upon their similarity with other entities that you know to be alive. Reflecting more deeply, you will recognize that these fossils have a set of patterns in common with these other living creatures: they are roughly symmetrical, bilaterally so (the left side is similar to the right); they have repeated segmentation, a feature shared with even humans (think of the spine as an expression of this segmentation). The exercise of finding these deep similarities is a search for homology: a deep convergence in biological structure that indicates a probably common origin.

Are you ever likely to be fooled by this search for pattern: might you mistakenly identify as formerly living, something that was purely inorganic (mineral crystals, for example?) This is likely to have been the case with the curious Allan Hills 84001 meteorite which contained a tiny bilaterally symmetrical, segmented, structure that was initially interpreted as a fossil. The exercise of comparing such objects with living structures reveals some of difficulties of defining life by means of purely physical comparisons. We need a more complex definition of life.

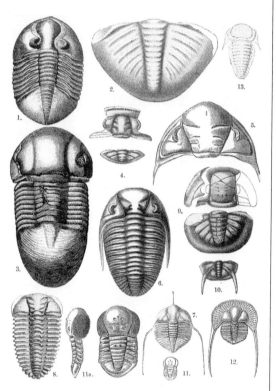

Lower Ordovician, example of typical forms of life. *Source:* Wellcome Images / Public Domain.

A Primer on Human Impacts on the Environment: The Conceptual Approach, First Edition. Liam Heneghan.
© 2023 John Wiley & Sons Ltd. Published 2023 by John Wiley & Sons Ltd.

Living things are shaped by their environment. In operationalizing the seemingly vague concept of "environment", ecologists identify those ecological factors that are especially influential in determining how living things flourish, and where, among many possible biogeographic settings, a given species is most likely to attain maximum fitness (that is, a level of reproductive success that can be evaluated over generations). The living entity that is placed at the center of a conceptual "mandala" can be an individual representative of a given population of just one species, though we can also ask questions about the environmental factors influencing an entire biotic community, or even the behavior of whole ecosystems.

It should be clear from our encounter with the concept of the mandala in the last chapter that the direction of influence between the living entity and its environment can flow in both directions. Factors in the environment influence the living entity and, in turn, the living entity modifies its environment. Such influences can be small in the case of individual organisms in local circumstances, but we can also ask the following question: what is the collective influence exerted by an entire population of a species, or, indeed, collectively of *all living things* on a planetary scale? In this, and the next chapter, we illustrate that the bidirectional flow of influence is not just an accidental feature of life, but may be its essential property: that is, life and environment are indissolubly linked. In particular, concepts that highlight how living things may influence the very conditions that determine the habitability of the planet are considered first, before we return to a theme that we initiated at the end of the previous chapter, that is, what sort of pressure do humans, in aggregate, assert on the planet. How does the impact of humanity in aggregate influence the ongoing habitability of the planet both for ourselves and the rest of nature?

7.1 An Environmental Definition of Life

Life is notoriously difficult to define [1]. Jean Gayon, a philosopher of biology, observes that although we harbor deeply-felt intuitive notions about what constitutes life, many scientists (and philosophers) are "skeptical about the possibility of defining life in a strong way..." [1]. Though the task of defining life is thus a nontrivial one – definitions tend to be either too specific for many purposes or so general as to be almost meaningless – yet a definition of life appropriate for environmental analysis must surely emphasize the connection between the living entity and the environment that surrounds it. An appropriate definition will then provide a basis for conceiving links between the regulatory actions of living beings and key environmental processes on a variety of scales from local to global.

Some definitions of life can apply quite successfully to all earthly life, but would they hold true for life beyond this planet where the challenge of identifying new life forms excites so much speculation? Though the criterion that the definition of life be applicable on planets or celestial objects other than our own, might seem over-exacting, nonetheless, it is useful, because it ensures that we are closing in on identifying the universal, or essential, features of life, and not ones that merely reflect the contingencies of the unique evolutionary history of our own world. Besides, as we will see, attempting to answer questions about life on other planets led to conceptual innovations that were fruitfully applied to Earth.

An example of a fairly robust definition that comes from an admittedly unlikely source is the definition provided by Friedrich Engels, the German political philosopher (and co-author with Karl Marx of *The Communist Manifesto* [1848]), who defines life as "the existence form of proteic structures, and this existence form consists essentially in the constant self-renewal of the chemical components of these structures. . ." [2]. Now, it's not especially clear why Engels alighted on "proteic structures" in particular as central to his definition, nor is it certain that we would find proteins to be an essential component of life elsewhere in the universe [2], but setting aside these eccentricities, the definition is useful in it generality (and, as we shall see, is close to one used in other areas of research).

The source of relevant environmental concepts can be surprising. The German philosopher, theorist, socialist and collaborator (and financial supporter) of Karl Marx, Friedrich Engels (1820–1895), wrote on an exceptionally broad range of topics. His application of revolutionary principles to science can be read in Dialectics of Nature, an unfinished 1883 manuscript, and posthumously published. There he provides a definition of life that has some striking similarities to the one used in the search for life elsewhere in the universe. *Source:* George Lester / Wikipedia Commons / Public Domain.

Some definitions may prove to be philosophically interesting, general enough to get at some universal properties of life, and yet might seem almost too capacious. Physiologist Claude Bernard (1813–1878), for example, stated the five general characteristics of life as follows: organization, generation, nutrition, development, and death. Depending upon the meaning of some of these terms ("development" and "death," in particular), this approach may admit some curious inorganic processes into the family of the living. Thus, in general, definitions that are formulated loosely, may apply to entities such as viruses and prions whose status as living beings is disputed, or indeed may blur the distinction between life and artificial life [3, 4]. Definitions that focus too purely on metabolism – transforming an energy source and producing waste – can potentially label fire as alive [5]!

So what might we expect from a universally applicable definition of life? Turning back to our discussion of the environmental mandala, you will recall that, in addition to its being reliant upon the environment, we focused on the capacity of an organism to transform its "medium" as a central implication of the ecological worldview. The universality of this state of affairs – that is, the mutuality of the living entity and its environment – derives, as we shall see, from the fact that the properties of living things must ultimately be consistent with the fundamental laws of thermodynamics (as must all entities and processes). A consideration of life in the context of these laws of thermodynamics – those fundamental laws that relate heat, work, temperature, and energy – will provide us access to the most general of all possible accounts of life. So let us examine, briefly, how definitions of life cohere with thermodynamic constraints.

Although there are three thermodynamic laws, it is the second law of thermodynamics that is most often invoked in discussions about living entities and their environment. The second law states that entropy, which can be thought of as a gauge of randomness in a closed system, must always spontaneously increase in the universe. Considering the orderliness of life and its evolutionary development, life seems to contradict, superficially at least, the edicts of the second law [6].

Therefore, there has been interest in reconciling the seeming incompatibility of life (as a highly ordered state, or a state of low entropy) with the second law. A definition of life that attempts this reconciliation will be seen as helpful, since our definition of life will then be consistent with fundamental physical laws and as a consequence will be applicable everywhere in the universe.

The physicist Erwin Schrödinger writing in his 1944 book, *What Is Life? The Physical Aspect of the Living Cell*, posed the problem as follows: "It is by avoiding the rapid decay into the inert state of 'equilibrium' [maximum randomness] that an organism appears so enigmatic; so much so, that from the earliest times of human thought some special nonphysical or supernatural force was claimed to be operative in the organism, and in some quarters this is still claimed. How does the living organism avoid decay?" [7]. The solution to this seeming paradox, Schrödinger pronounced, is found in "the device by which an organism maintains itself stationary at a fairly high level of the orderliness...[and this] really consists in continually sucking orderliness from its environment" [7]. Thus, living things maintain their low state of entropy (high order) by creating disorder all around them. A disordered and dynamic environment becomes a signature of life.

Though, as we have seen, no definition of life will be universally agreed upon, nonetheless, a definition is a good one if it alights upon some aspect of the essence of life, while also suiting the particular purpose to which it is applied. Though the definition used by the National Aeronautics and Space Administration (NASA), for example, doesn't explicitly reference the second law of thermodynamics, it clearly gets at this fundamental property of life and its definition is well suited to the particular purpose it must serve. In this instance, the purpose specifically allows for the identification of extraterrestrial life. The NASA definition is as follows, "life is a self-sustained chemical system capable of undergoing Darwinian evolution" [8, 9]. We should note, that with the exception of it making a reference to proteic structures, the NASA definition is not too far removed from that of Engels. If one felt compelled to search for life in strange places, this definition suggests an approach and sets up reasonable expectations for an investigation. Furthermore, the NASA definition may serve as a general environmental definition of life. This is because the processes required for a "self-sustaining chemical system" necessarily involve drawing upon the resources of the environment (a *source*) in which that living thing is found while recognizing that the living thing must, in turn, emit wastes into the environment (a *sink*). All living beings, earth-bound or extraterrestrial, are enmeshed in their environments and transform it in distinctive ways. In addition, the criterion of being capable of undergoing Darwinian evolution implicates a biotic individual in the "struggle for life." Living beings that are capable of undergoing Darwinian evolution exist, by axiom, in populations competing for resources (for it is resources, in Darwin's account, that living beings struggle over). Those characteristics of the individual that provide greatest aid in this struggle can be passed on to descendants (if these characteristics are inheritable). Though Darwin's principles were formulated to account for the diversity of life on this planet, they are universal and can therefore be applied to the complex adaptive systems of life anywhere. Some authors have advocated for leaving off the Darwinian codicil to the NASA definition, arguing that it places too stringent a criterion on the task of finding life. The definition would thereby be transformed into the following: life is "a system which is self-sustaining by utilizing external energy/nutrients owing to its internal process of component production..." [2].

With or without the insistence that life conforms to Darwinian struggles, this style of definition is useful for it insists that a metabolic transformation of an environment is a key characteristic of life. Therefore, detecting life involves investigating diagnostic environmental transformations: either uncovering the exploitation of resources to maintain otherwise improbably ordered structures, or, more commonly, detecting recognizable metabolic wastes [10, 11].

7.1.1 The Search for Life on Mars

A brief example of how the search for life on planets other than our own has been conducted can help in clarifying the cogency of our definitions of life. NASA's Viking probes, launched in 1975 on a mission that included the search for life on Mars, illustrate the application of an environmental approach to the identification of vital signs. One of the experiments on board the Viking spacecraft tested if a soil sample when supplied with a radioactive ^{14}C isotope (the source) would produce detectable radioactive gases released into a sealed experimental chamber (the sink). Though the initial results were reported as positive for signs of possible microbial life, these findings have remained controversial [12, 13]. Setting aside the equivocality of the results, metabolic approaches to the identification of life remain the most tractable means of approaching the question of whether a planet is hospitable to living forms. The approach is based upon the assumption that when a living entity is present it will be "self-sustained" by drawing from its environment and leaving an environmental trace.

In this spirit, the recent detection of phosphine in the atmosphere of Venus generated considerable interest. On earth phosphine is associated with either anthropogenic activity or with a microbial presence; therefore it makes a good candidate as a biosignature [14]. This discovery in the atmosphere of Venus was considered so unexpected that it has led to much speculation about its origin; these speculations include the possibility that there is life on that planet [14, 15]. Such findings illustrate, on the one hand, the attractiveness of a metabolic/environmental approach to the detection of life, and yet, on the other hand, if, prompted by these observations, we discover that phosphine may be produced by previously unknown photo- or geochemistry, it underscores the concern that overly generalized definitions of life may admit into the family of the living some very exotic members [16]. So, we must proceed with caution in our search for life on other parts of the solar system and beyond.

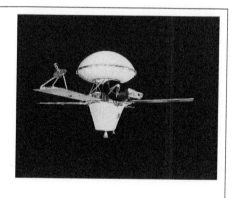

The Viking spacecraft was designed to conduct a range of observations on Mars, including investigating the possibility of life in Martian soils (NASA/Wikipedia Commons/ Public Domain). Though the result were ambiguous, the test for life was based upon a recognition of the indissolubility of life and environment. *Source:* NASA / Wikipedia Commons / Public Domain.

As it is on Mars (and Venus), and presumably elsewhere in the universe, so it must be on Earth: any living entity capable of self-production and self-maintenance through adaptive interactions with a medium, will produce a change in that medium or environment [17].

In spite of the difficulties in defining life it should be clear from the foregoing that, firstly, living beings on this or any other planet must conform to the fundamental laws of the universe – no special forces or exemptions need to be invoked – and, secondly, that the relationship between the organism and its environment is a reciprocating one. Though the general principle that organisms and their environment needs to be studied in tandem, nonetheless, the *scale* upon which the transformation of environments by living beings occurs can be quite various.

Individual organisms may affect environmental transformations in quite modest ways; on the other hand, it is possible that collectively an entire species – as humans are conjectured to do – may transform a region or indeed an entire planet. Let us first turn now to examining these local transformations.

7.2 The Transformation of Local Environments

Living things as we have established affect the transformation of the environment in two ways. Firstly, they must extract resources so as to maintain the "nonrandomness" (sensu Schrödinger) of their internal structures, and, secondly, they must necessarily expel metabolic waste into the "medium" that they inhabit. This distinction between exploiting sources and befouling sinks resonates, as we have seen, with the basic environmental model presented by economist Herman Daly [18].

The depletion of resources and the fouling of surroundings can occur on very local scales. In calling attention to the struggle for scarce resources in local environments, Darwin made it a central axiom of his concept of natural selection. In *On the Origin of Species* (1859) Darwin wrote ". . .the struggle almost invariably will be most severe between the individuals of the same species, for they frequent the same districts, require the same food, and are exposed to the same dangers" [19]. That is, two individuals of the same species will compete most intensely, and locally, in the struggle for existence because they have roughly the same resource needs. Though this point is made clearly enough in Darwin's writing, a more profoundly developed concept of competition for resources between species in the arena of local habitats became a cornerstone of twentieth century community ecology [20]. In plant communities, neighboring individuals belonging to different species are fixed in place. Individuals from a species that can survive at the lowest resource level (for example phosphorus in soil) can, under some circumstances, displace less able competitors [21, 22]. Thus, the manner in which a living thing – in this example, the plant – can modify its environment in the struggle for life is consequential in determining the local presence and absence of species. Success at garnering scarce resources means that there is less for poorly performing competitors.

Animals (including humans), in contrast to plants, exploit an environment in which resources – like food – are patchily distributed on a wider variety of scales [23]. If one resource patch is in danger of being exhausted the animal can move to another patch. In some circumstances they return when resources are replenished [24]. Though human foraging can be understood using the same ecological framework as is applied to other animals, the patterns of resource exploitation by human communities can be very complex, and there is not always a good fit between the way humans use resources and the predictions from ecological theory [25–28]. But, as is true for other organisms, when humans appropriate a greater share of resources, fewer remain for other species (though, of course, the complication with respect to human use is that humans will typically cultivate their own resources, through farming and manufacturing).

> The *carrying capacity* of a particular environment is the maximum number of organisms of a given species that the environment can support without an erosion of habitat quality.

The aggregate resource capacity of a region and its ability to support populations is the basis for estimates of an "environmental carrying capacity" [29]. Carrying capacity is defined in relation to a maximum population that can be supported indefinitely in a given place (although other definitions abound) [30, 31]. The application of the concept to human populations – by asking how many people can the resources of the planet support – has proved both difficult but also controversial, although at its core the concept embeds the incontestable notion that all organisms, humans included, are in a dynamic relationship with the resources that sustain them [32–34]. The carrying capacity in its general conceptual form depicts how much of the resource base is available for use in an environment without exhausting that environment's regenerative capacity. To put it another way, under what circumstances can a population transform a local environmental without impairing it for future generations?

As well as utilizing (and potentially depleting) resources in their environment – that is, transforming the source of resources – all organisms export metabolic waste into their environment. Animal waste, in the form of both the products of egestion (feces, and so on) or excretion (often, but not exclusively, nitrogenous waste) alter the environments that receive them. Animal waste can provide some resources upon which other organisms rely, and the processing of this waste is integral in the cycling of several nutrients [35, 36]. Such waste can, however, be toxic in the environment and many (though not all) organisms avoid contact with their own and other animals' excretory waste [37].

Humans are not exceptional in their ability to transform environments on a local scale – as should be abundantly clear from what we have already discussed. By means of engaging with a local environment, we can shape the nature of our local resource base and our wastes can be environmentally significant both as a resource for some decomposer organisms but may also act as a pollutant. There are many case studies from environmental research on the ability of *premodern* communities to shape their local habitat. This is important to acknowledge, since it is sometimes erroneously assumed that alterations of environment by humans is exclusively a contemporary phenomenon.

A few examples of premodern humans exhausting resources and polluting sinks will suffice to illustrate this point.

1) In reexamining the causes of *environmental collapse on Easter Island*, the ecologist Jared Diamond (who we will meet again in Chapter 11) noted that by the 1700s Easter Island almost completely lacked trees over 3 m tall. However, pollen analysis in swamp cores revealed the former prevalence of a giant palm tree and other species that no longer existed when European populations arrived on the island. Indeed, careful examination of wood remains in this analysis discovered 20 other tree species exterminated by settlers. Trees were gone by 1650 CE. Controversially, Diamond concluded that deforestation coupled with changes in horticultural practices, alongside some other factors – possible drought, the introduction of invasive species, all occurring in the context of general environmental fragility – resulted in a collapse of the Polynesian settlers of the island [38, 39].

2) Ecologists who have examined *the prehistoric and preindustrial modification of Europe* argue that the clearing of forests to establish cropland and pasture, and the exploitation of forests for fuel wood and construction materials, was enormously consequential ecologically. Although the de-vegetation of Europe was a significant regional transformation, it also may have resulted in previously unrecognized impacts on a global scale. In their simulation of extensive European deforestation that had occurred by 1000 BCE, paleoecologist Jed O. Kaplan and his colleagues reported that the deforestation-related perturbation of the carbon cycle during the Holocene era has been greatly underestimated, and was significant enough to be translated into the relatively early human impact on the climate system [40].

3) The hydrologist Marc Leblanc and his colleagues investigated *mining pollution in southwestern Spain* dating back to 4500-years ago. They showed that mining occurred early (Copper Age) in the Rio Tinto area, and that this resulted in watershed-scaled metal contamination. These metals accumulated and were immobilized in sediments found in the adjacent estuary [41].

4) In a similar fashion, Fiona Roos-Barraclough and her colleagues investigated *the accumulation of mercury in peat over a 14500-year period in the Swiss Jura Mountains.* Although some of the mercury can be traced to volcanic sources, they revealed evidence of anthropogenic input. This research demonstrated that even though mercury accumulation indicative of pollution prevailed especially during the industrial period it also occurred in a less pronounced way in pre-industrial times [42].

5) Finally, using a palynological approach (this approach included studying dinoflagellate cysts – these are the remains of single-celled algal relatives – as well as pollen and spores) paleo-ecologists Francesca Sangiorgi, and T.H. Donders reconstructed *150 years of nutrient accumulation in the Italian north-western Adriatic Sea.* Especially after the early twentieth century, this accumulation of nutrients resulted from increased wetland reclamation and forest clearance [43].

7.3 Can Living Things Transform the Globe? An Introduction to the Gaia Hypothesis

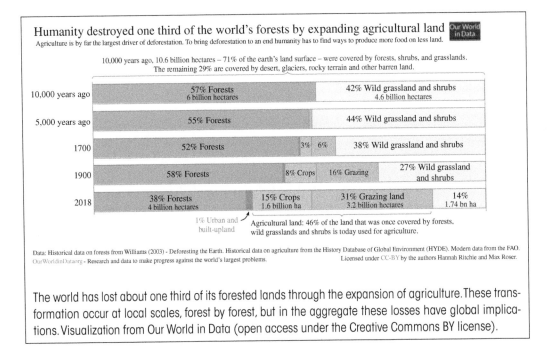

Humanity destroyed one third of the world's forests by expanding agricultural land

Agriculture is by far the largest driver of deforestation. To bring deforestation to an end humanity has to find ways to produce more food on less land.

10,000 years ago, 10.6 billion hectares – 71% of the earth's land surface – were covered by forests, shrubs, and grasslands. The remaining 29% are covered by desert, glaciers, rocky terrain and other barren land.

Data: Historical data on forests from Williams (2003) - Deforesting the Earth. Historical data on agriculture from the History Database of Global Environment (HYDE). Modern data from the FAO. OurWorldinData.org - Research and data to make progress against the world's largest problems. Licensed under CC-BY by the authors Hannah Ritchie and Max Roser.

The world has lost about one third of its forested lands through the expansion of agriculture. These transformation occur at local scales, forest by forest, but in the aggregate these losses have global implications. Visualization from Our World in Data (open access under the Creative Commons BY license).

It is inarguable that large regional swaths of the globe have been transformed by human action through the accumulation of a persistent and substantial alteration of local environments. In what follows we are broadening our focus to include all living things, and not just humans. Our guiding question is this: can the effects of living things register an impact beyond the local or regional scale? The objective of this section is to show that there are good reasons for accepting that living things – certain functional types, as well as all living things considered collectively – can transform, not just regions, but the entire global environment. The objective in showing that non-human organisms acting in local arenas can cumulatively affect global processes should not be taken as absolving humans of our responsibility for functional modifications at a global scale (when these modifications have negative consequences), but rather to show that the notion of living entities manifesting impacts on global processes is not conceptually anomalous. In fact, this might be expected considering the indissoluble link between all living things, humans included, and the environment.

There are some relatively celebrated incidents where the metabolic waste of nonhuman organisms are known to have transformed the world. The emission of oxygen, produced as photosynthetic "waste," for example, proved dramatically important in shaping the environment for other organisms. Habitats that were originally iron rich and relatively oxygen poor were replaced by oxygen-rich environments. These oxygen rich environments, ultimately, became hospitable to complex multicellular life [44]. The transformation of local environments into ones with a high concentration of oxygen after the emergence of the first photosynthetic cells thus resulted in new and possibly profound selective pressure on anaerobic microbes, that may have diminished the extent of their available habitat [45, 46]. The subsequent history of life depended to a surprising extent upon this "pollution" event.

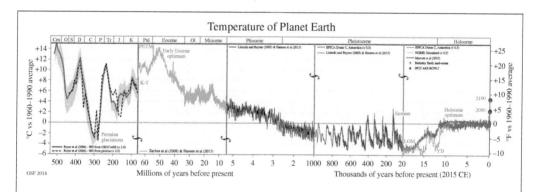

Global average temperature estimates for the last 540 My. The remarkable constancy of the global temperatures over very long periods of time suggest the existence a regulatory mechanism. Planetary engineer James Lovelock suggests (controversially) that the temperature of the globe is maintained via feedback mechanisms from the biosphere. This the essence of the Gaia hypothesis.

James Lovelock (1919–2022), an engineer and environmental scientist, formulated and popularized the idea that the biosphere transforms and regulates the abiotic environment of the planet. This idea of the global environment being tightly regulated by life, is called the Gaia hypothesis (or theory). It may not be surprising that a point of departure for Lovelock's ideas, as we will see, came from his frustration over certain attempts to define life. Like many scientists who have sought a definition of life, Lovelock, conceded that the "state of life...has so far resisted all attempts at a formal physical definition" [16]. In his provocative (and still very readable) book, *Gaia: A New Look at Life on Earth* (1979), Lovelock comments wryly that, "most forms of life can be instantly recognized without prior experience and are frequently edible." Yet a central argument of his book engages more substantially with the problem of detecting life. Lovelock shares a revealing anecdote about being asked his opinion by colleagues at the Jet Propulsion Lab at the California Institute for Technology on how he would detect life on Mars. At that time, his colleagues were involved in the design of the microbial tests that would be taken aboard the Viking probes; Lovelock had seemed skeptical. Responding to queries about potential alternatives to the proposed tests for life, Lovelock suggested that the team look for entropy reduction on Mars (that is, they should ask how life had disordered the thin Martian atmosphere), since "this [disordering] must be a general characteristic of all forms of life." (This suggestion, you will have noticed, is just the implications of Schrödinger's thinking writ large.) Upon deeper reflection, Lovelock proposed a simple test: examine disequilibrium among reactive atmospheric gases as evidence of life. One could simply use an infrared telescope or other astronomical techniques to detect the presence of a planetary biosphere based upon the presence of anomalous concentrations of gases that should be in equilibrium with the environment (equilibrium in this context means that all the chemical reactions that might have occurred will have occurred). If the atmosphere is in chemical equilibrium there is no current life on that planet [47]. If there is life on a planet, the atmosphere would be in disequilibrium since living things would be continually disordering it (as is the case on Earth). Evidence that any planet may harbor life can be sought by comparing the atmospheric elements at equilibrium versus what is observed when the atmosphere is regulated by life [16, 48].

Cybernetics: This term was coined by mathematician Norbert Wiener to refer to systems of control and communication in animals and machines. Cybernetic systems are those where there is tight regulation of the states of the system – like the temperature in a room regulated by a thermostat – thorough feedback cycles. This sort of feedback can be referred to as homeostatic (deriving from Greek roots meaning standing still).

Although, Lovelock's approach to the life detection of life on Mars was not subsequently employed in NASA's exobiological work – they chose the small chambered radiolabel experiments discussed above – the real significance of his ruminations about life's diagnostic features emerged when he applied these ideas to understanding the functioning of the Earth as a complex adaptive system. Baldly stated, the Gaia hypothesis, as formulated by Lovelock (and his collaborator, biologist Lynne Margulis), is that "the climate and chemical properties of the Earth now and throughout history seem always to have been optimal for life" and that, importantly, these physical and chemical conditions have been "actively made fit and comfortable by the presence of life itself" [49, 50]. Planetary conditions are made by and for the benefit of the biosphere. This curious state of affairs should not be expected by chance – living things, Lovelock concluded, must modify the atmosphere to maintain conditions (including climate) in ways that maintain planetary suitability of life.

Lovelock's audacious claim has generated considerable debate [51–53]. At least some of the debate has focused upon the supposed incompatibility between claims that Gaia can cybernetically self-regulate (and, in some accounts at least Gaia is seen as a sort of super-organism) and the tenets of Darwinian natural selection [54]. How can the idea of organisms blindly struggling to maximize their evolutionary fitness (a supposedly "selfish" process) be reconciled with the idea of a seemingly benign self-regulating Gaian system [52, 55]? Lovelock and colleagues have attempted to reconcile this tension by illustrating how the growth dynamics of "model" plants in a mathematical simulation they called "Daisyworld" can result, in simulations at least, in the homeostatic regulation of the temperature of the earth [56]. It is not at all clear that these very simple models have fully resolved the difficulties for this aspect of Lovelock's Gaia hypothesis, but one thing is apparent: it provoked significant interest in exploring the role of organisms, other than humans, in changing biogeochemistry on a global scale [53, 57, 58]. As Watson and Lovelock wrote in 1983 "on earth, modification of the environment by living things is apparent on any scale that one cares to look at, *up to and including the global scale*" [emphasis, mine.] [56].

Stripped of its claims that the biosphere imposes regulation upon planetary conditions through homeostatic regulation, the Gaia hypothesis stills serves to underscore the reality that biological processes can exert a large-scale influence on geochemical cycles. In doing so it adds insights into those processes that couple biotic and physical systems on Earth [59]. "Gaia-lite" (as I call the Gaia hypothesis relieved of its implications about long-term homeostasis) still makes significant and testable claims [53].

Three examples of important discovers linking biota to globally significant, though obscure, transformations of the environment that were predicted by Gaia theory are:

1) When our actual living planet is compared against the theoretical steady state (or equilibrium) conditions in the absence of life, these two states are vastly different. Our atmosphere, as Lovelock observed, exists in a state of disequilibrium. This "paired planet" approach reveals the extent of the regulatory control exercised by life over the physics and chemistry of the planet. It is seen in the cycling of the major nutrients (carbon, nitrogen, oxygen, phosphorus, sulfur and so on) [53].

2) Lovelock predicted that there is a biological transfer of selenium from the ocean to the land in the form of dimethyl selenide [60]. We now know that a major pathway for the production of gaseous selenium occurs in surface ocean waters; phytoplankton have a role to play in the emission of selenium to the atmosphere. This important micronutrient is supplied to the terrestrial environment from marine environments [61].

3) Another prediction of Gaian theory is that there should be a transfer of elements via biogenic gases from ocean to land. Indeed, the oceans appear to be a major source of dimethyl sulfide cloud-condensation nuclei (CCN). Planktonic algae produce these nuclei and Lovelock and others have proposed that this production may exert an influence over the climate [62].

Discussions about the transformative effects that the biosphere can have on the global environment are not merely academic exercises. Lovelock, like many subsequent Gaian theorists, has been interested in examining the role of human pollution on planetary functioning. Are humans in danger of overwhelming the regulation of planetary feedback necessary for the stability of climate, and other vital global ecological functions?

In *Gaia: A New Look at Life on Earth*, Lovelock expressed skepticism about human's ability to unleash anything as destructive on Earth greater than what nature is already capable of: "there is no Geneva Convention to limit natural dirty tricks" [16]. The "dirty tricks" that Lovelock had in mind include the release of toxins, aflatoxin released by *Aspergillus* molds, the polypeptides produced by the Death-cap fungus, and the whole lethal arsenal of chemicals produced in the natural world. He makes the case that the toxicity produced by human pollution has to some extent a "natural background." Although there may be circumstances where human-made pollution can be problematic, and Lovelock did not completely downplay the possible consequences of increases in atmosphere CO_2, or ozone depletion and at the time he published that book – that is, 1979 – he, nonetheless struck a cautious tone, writing that we are guilty of "ignorance of planetary control systems." As we shall see, his views on this issue were subsequently to change quite dramatically.

7.4   From Gaia to the Anthropocene

In the preceding pages, we discussed the circumstances whereby resources are depleted by living things and waste is produced as a consequence. Resource uptake and waste production are the inevitable outcomes of living processes and both uptake and waste are incorporated into environmentally appropriate definitions of life.

Like other living things, humans transform the environment. Local impacts occupy much of the attention of environmental policy makers [63, 64]. However, the sorts of problems that transform the environmental sciences into "crisis disciplines," that is, the disciplines dealing with matters of "acute concern," for the planet's future, are those problems that manifest not at a local scale but, rather those occurring at a the scale of the globe [65, 66]. There is a growing awareness that these global environmental impacts are now consequential enough to transform life on the planet not only for humans but by most living things.

By 2007, Lovelock had dramatically reevaluated assessments that he made in the 1970s concerning the potential of humans to overwhelm Gaia's regulation of planetary systems. In later years he argued that it is "already too late, to prevent the global climate from 'flipping' into an entirely new equilibrium state that will leave the tropics uninhabitable, and force migration to the poles" [67]. This dramatic elevation in concern expressed by a leading (albeit provocative) scientist reflected a growing sense of disquiet in environmental disciplines that the human transformation of the global had reached levels where extremely undesirable impacts on planetary systems would be felt.

That humans have the capacity to transform the globe is captured by the neologism "the Anthropocene" [68, 69]. The Anthropocene encapsulates the conjecture that human action on a planetary scale is now so momentous that we will leave not merely a temporary mark upon the planet, but rather our reshaping of the plantet will impressed upon the geological record. Understanding this conjecture will form the main topic of the next chapter.

7.5   Coda: From Mandala to Human-Dominated Planet

This chapter asserted that fundamental to any definition of life useful for environmental analysis is a recognition of a link between a living entity and its environment. A living being generates disorder in its surroundings; it cannot avoid doing so. Environmental disordering can range from the modest and local to the prodigious and global. Before we are able to assess claims that humans are asserting an overwhelming impact at a global scale, in this chapter we examined in some detail the controversial idea that life, collectively, has played a role in regulating aspects of the chemistry of the atmosphere, and by doing so can serve to regulate the temperature of the globe, maintaining it at a level required for sustaining life. In the next chapter, we examine some direct precursors to the concept of the Anthropocene: a designation of a new geological era defined by human influence. We ask if the scale and severity of the global human impact has now breached the limit of the planet's ability to provide the key resources for our economies as well as impairing its capacity to serve as a sink for our wastes.

References

1 Gayon, J. (2010). Defining life: synthesis and conclusions. *Origins of Life and Evolution of Biospheres* 40 (2): 231–244.

2 Luisi, P.L. (1998). About various definitions of life. *Origins of Life and Evolution of the Biosphere* 28 (4–6): 613–622.

3 Villarreal, L.P. (2004). Are viruses alive? *Scientific American* 291 (6): 100–105.

4 Langton, C.G. (1986). Studying artificial life with cellular automata. *Physica D: Nonlinear Phenomena* 22 (1–3): 120–149.

5 Davies, P.C.W. and Davies, P. (2000). *The Fifth Miracle: The Search for the Origin and Meaning of Life*. Simon and Schuster.

6 Monod, J. (1971). *Chance and Necessity: An Essay on the Natural Philosophy of Modern Biology*. New York: Alfred A. Knopf.

7 Schrodinger, E. (2012). *What Is Life?: With Mind and Matter and Autobiographical Sketches*. Cambridge University Press.

8 Horowitz, N. and Miller, S.L. (1962). *Current Theories on the Origin of Life*. Fortschritte der Chemie Organischer Naturstoffe/Progress in the Chemistry of Organic Natural Products/Progrès Dans la Chimie Des Substances Organiques Naturelles, 423–459. Springer.

9 Deamer, D.W. and Fleischaker, G.R. (1994). *Origins of Life: The Central Concepts*. Jones & Bartlett Pub.

10 Levin, G.V. and Straat, P.A. (2016). The case for extant life on Mars and its possible detection by the Viking labeled release experiment. *Astrobiology* 16 (10): 798–810.

11 Navarro-Gonzalez, R., Vargas, E., de la Rosa, J. et al. (2011). Reanalysis of the Viking results suggests perchlorate and organics at midlatitudes on Mars (vol. 115, E12010, 2010). *Journal of Geophysical Research-Planets* 115 (E12): https://doi.org/10.1029/2010je003599.

12 Guzman, M., McKay, C.P., Quinn, R.C. et al. (2018). Identification of chlorobenzene in the Viking gas chromatograph-mass spectrometer data sets: reanalysis of Viking mission data consistent with aromatic organic compounds on Mars. *Journal of Geophysical Research: Planets* 123 (7): 1674–1683.

13 Levin, G.V. (1972). Detection of metabolically produced labeled gas: the Viking Mars Lander. *Icarus* 16 (1): 153–166.

14 Greaves, J.S., Richards, A.M., Bains, W. et al. (2020). Phosphine gas in the cloud decks of Venus. *Nature Astronomy* 5: 1–10.

15 Bains, W., Petkowski, J.J., Seager, S. et al. (2021). Phosphine on Venus cannot be explained by conventional processes. *Astrobiology* 21 (10): 1277–1304.

16 Lovelock, J.E. (1979). *Gaia: A New Look at Life on Earth*. Oxford Paperbacks.

17 Damiano, L. and Luisi, P.L. (2010). Towards an autopoietic redefinition of life. *Origins of Life and Evolution of Biospheres* 40 (2): 145–149.

18 Daly, H.E. (1996). *Beyond Growth: The Economics of Sustainable Development*. Beacon Press.

19 Darwin, C. (1996). *The Origin of Species: Oxford World's Classics*. Oxford University Press.

20 Giller, P. (2012). *Community Structure and the Niche*. Springer Science & Business Media.

21 Tilman, D. (1985). The resource-ratio hypothesis of plant succession. *The American Naturalist* 125 (6): 827–852.

22 Tilman, D. (1982). *Resource Competition and Community Structure*. Princeton University Press.

23 Clark, G. (1994). *Space, Time and Man: A Prehistorian's View*. Cambridge University Press.

24 Beecham, J. and Farnsworth, K. (1998). Animal foraging from an individual perspective: an object orientated model. *Ecological Modelling* 113 (1–3): 141–156.

25 Cashdan, E., Barnard, A., Bicchieri, M. et al. (1983). Territoriality among human foragers: ecological models and an application to four bushman groups [and comments and reply]. *Current Anthropology* 24 (1): 47–66.

26 Smith, E.A., Bettinger, R.L., Bishop, C.A. et al. (1983). Anthropological applications of optimal foraging theory: a critical review [and comments and reply]. *Current Anthropology* 24 (5): 625–651.

27 Cronon, W. (2011). *Changes in the Land: Indians, Colonists, and the Ecology of New England*. Hill and Wang.

28 Johnson, A.W. and Earle, T.K. (2000). *The Evolution of Human Societies: From Foraging Group to Agrarian State*. Stanford University Press.

29 Young, V.A. (1938). The carrying capacity of big game range. *The Journal of Wildlife Management* 2 (3): 131–134.

30 Kirchner J., Ledec, G., Goodland, R.J., and Drake, J.M. (1985). Carrying capacity, population growth, and sustainable development. Rapid population growth and human carrying capacity: Two perspectives. World Bank Staff Working Papers.

31 Cohen, J.E. (1996). *How Many People Can the Earth Support?* WW Norton & Company.

32 Cohen, J.E. (1995). Population growth and earth's human carrying capacity. *Science* 269 (5222): 341–346.

33 Dhondt, A.A. (1988). Carrying capacity: a confusing concept. *Acta Oecologica* 9 (4): 337–346.

34 Postel, S. (1994). Carrying capacity: Earth's bottom line. *Challenge* 37 (2): 4–12.

35 Sprent, J.I. (1987). *The Ecology of the Nitrogen Cycle*. Cambridge University Press.

36 Stevenson, F.J. and Cole, M.A. (1999). *Cycles of Soils: Carbon, Nitrogen, Phosphorus, Sulfur, Micronutrients*. Wiley.

37 Hatch, A.C., Belden, L.K., Scheessele, E., and Blaustein, A.R. (2001). Juvenile amphibians do not avoid potentially lethal levels of urea on soil substrate. *Environmental Toxicology and Chemistry: An International Journal* 20 (10): 2328–2335.

38 Diamond, J. (2007). Easter Island revisited. *Science* 317 (5845): 1692–1694.

39 Merico, A. (2017). Models of Easter Island human-resource dynamics: advances and gaps. *Frontiers in Ecology and Evolution* 5: 154.

40 Kaplan, J.O., Krumhardt, K.M., and Zimmermann, N. (2009). The prehistoric and preindustrial deforestation of Europe. *Quaternary Science Reviews* 28 (27–28): 3016–3034.

41 Leblanc, M., Morales, J.A., Borrego, J., and Elbaz-Poulichet, F. (2000). 4,500-year-old mining pollution in southwestern Spain: long-term implications for modern mining pollution. *Economic Geology and the Bulletin of the Society of Economic Geologists* 95 (3): 655–661.

42 Roos-Barraclough, F., Martinez-Cortizas, A., Garcia-Rodeja, E., and Shotyk, W. (2002). A 14 500 year record of the accumulation of atmospheric mercury in peat: volcanic signals, anthropogenic influences and a correlation to bromine accumulation. *Earth and Planetary Science Letters* 202 (2): 435–451.

43 Sangiorgi, F. and Donders, T.H. (2004). Reconstructing 150 years of eutrophication in the North-Western Adriatic Sea (Italy) using dinoflagellate cysts, pollen and spores. *Estuarine, Coastal, and Shelf Science* 60 (1): 69–79.

44 Sperling, E.A., Knoll, A.H., and Girguis, P.R. (2015). The ecological physiology of Earth's second oxygen revolution. *Annual Review of Ecology, Evolution, and Systematics* 46: 215–235.

45 Bilinski, T. (1991). Oxygen toxicity and microbial evolution. *Biosystems* 24 (4): 305–312.

46 Fenchel, T. and Finlay, B. (2008). Oxygen and the spatial structure of microbial communities. *Biological Reviews* 83 (4): 553–569.

47 Lovelock, J.E. (1975). Thermodynamics and the recognition of alien biospheres. *Proceedings of the Royal Society of London Series B Biological Sciences* 189 (1095): 167–181.

48 Lovelock, J. (2000). *The Ages of Gaia: A Biography of our Living Earth*. USA: Oxford University Press.

49 Lovelock, J.E. and Margulis, L. (1974). Atmospheric homeostasis by and for the biosphere: the Gaia hypothesis. *Tellus* 26 (1–2): 2–10.

50 Margulis, L. and Lovelock, J.E. (1974). Biological modulation of the Earth's atmosphere. *Icarus* 21 (4): 471–489.

51 Kirchner, J.W. (2002). The Gaia hypothesis: fact, theory, and wishful thinking. *Climatic Change* 52 (4): 391–408.

52 Lenton, T.M. (1998). Gaia and natural selection. *Nature* 394 (6692): 439–447.

53 Volk, T. (2012). *Gaia's Body: Toward a Physiology of Earth*. Springer Science & Business Media.

54 Dawkins, R. (1982). *The Extended Phenotype*. Oxford University Press.

55 Doolittle, W.F. (2014). Natural selection through survival alone, and the possibility of Gaia. *Biology & Philosophy* 29 (3): 415–423.

56 Watson, A.J. and Lovelock, J.E. (1983). Biological homeostasis of the global environment: the parable of Daisyworld. *Tellus B: Chemical and Physical Meteorology* 35 (4): 284–289.

57 Dagg, J.L. (2002). Unconventional bed mates: Gaia and the selfish gene. *Oikos* 96 (1): 182–186.

58 Wood, A.J. and Coe, J.B. (2007). A fitness based analysis of Daisyworld. *Journal of Theoretical Biology* 249 (2): 190–197.

59 Kirchner, J.W. (1989). The Gaia hypothesis: can it be tested? *Reviews of Geophysics* 27 (2): 223–235.

60 Lovelock, J. (2010). *The Vanishing Face of Gaia: A Final Warning*. Basic Books.

61 Amouroux, D., Liss, P.S., Tessier, E. et al. (2001). Role of oceans as biogenic sources of selenium. *Earth and Planetary Science Letters* 189 (3–4): 277–283.

62 Lovelock, J. (1997). A geophysiologist's thoughts on the natural sulphur cycle. *Philosophical Transactions of the Royal Society of London Series B: Biological Sciences* 352 (1350): 143–147.

63 Gibbs, D. and Jonas, A.E.G. (2000). Governance and regulation in local environmental policy: the utility of a regime approach. *Geoforum* 31 (3): 299–313.

64 Press, D. (1998). Local environmental policy capacity: a framework for research. *Natural Resources Journal* 38 (1): 29–52.

65 Cox, R. (2007). Nature's "Crisis Disciplines": does environmental communication have an ethical duty? *Environmental Communication-a Journal of Nature and Culture* 1 (1): 5–20.

66 Bowler, P.J. (1993). *The Norton History of the Environmental Sciences*. WW Norton & Company.

67 Lovelock, J. (2007). The revenge of Gaia: earth's climate crisis & the fate of humanity. *Basic Books*.

68 Crutzen, P.J. (2016). Geology of mankind. In: *A Pioneer on Atmospheric Chemistry and Climate Change in the Anthropocene* (ed. P.J. Crutzen), 211–215. Springer.

69 Lewis, S.L. and Maslin, M.A. (2015). Defining the anthropocene. *Nature* 519 (7542): 171–180.

8

Gaia, the Noösphere, and the Anthropocene

Teleology is the examination of phenomena that can be best explained in terms of intention, design, or purposiveness.

The living parts of Planet Earth, considered as a system, are both *dependent upon* and *influential for* inanimate processes. In the last chapter, we discussed how these dependencies and influences result in the environmental transformation of planetary sources, sinks, and abiotic processes on a variety of scales. These scales range from the local up to and including the entire globe. James Lovelock's and Lynne Margulis's concept of Gaia crystalized the notion that reciprocal patterns of interconnection between the life and geochemical processes occur and are influential on a planetary scale [2, 3]. Stripped of the teleological implications found in at least some Gaian writing – that is, the suggestion that the processes are regulated with a particular purpose in mind – the idea that organisms, hydrological systems, atmosphere, and lithosphere (rocks and soil) are mutually influential is now a commonplace in the environmental sciences.

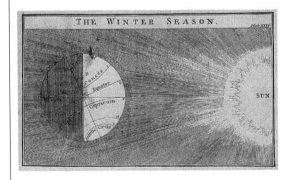

There are reasons for supposing that life, even intelligent life, flourishes throughout the universe, yet ours in the only living planet that we know of. Earth does not merely harbor life, rather it is a *living planet* in the sense that life both thrives on the planet and is influenced by the climatic and biogeochemical factors of this world, but it also, in turn, exerts a strong influence on those conditions.

The evolution of humans – shaped by the same processes that produced all species – represents a departure: no single species seems to have been quite so influential on the conditions of the planet. Should humans, therefore, be considered anomalous, and outside of nature, or simply as one in a long series of unusual events in the history of this four and a half billion year old planet?

Meteorology: a view of the Earth and the sun during winter (in the Northern hemisphere). Engraving after B. Martin. *Source:* Wellcome Images / Public Domain.

A Primer on Human Impacts on the Environment: The Conceptual Approach, First Edition. Liam Heneghan.
© 2023 John Wiley & Sons Ltd. Published 2023 by John Wiley & Sons Ltd.

The Gaia hypothesis is conceptually important; not only is it of inherent interest for environmental disciplines, but also as a point of departure for a significant body of research [4]. In addition, Gaia provides an important precedent for the idea that life can exert a planetary influence. In a complex system, oftentimes multiple elements at lower ecological scales are constrained by upper level holons; however, in certain circumstances, and Gaian processes seem to be among such circumstances, the collective pressure of the "parts" can dictate the stability of the whole. The notion that human beings, as one species among millions, can exert an overwhelming impact on planetary processes is a big claim, but it can hardly be seen as ecologically aberrant since there are precedents of other living entities regulating planetary affairs.

In what follows we will explore a range of related ideas – most of them older than the concept of Gaia – that also ground the idea that a living entity can exert such an influence. We will conclude with an account of why the notion of the Anthropocene is not, on the one hand, beyond the conceptual mainstream for environmental science, and, why, on the other hand, this idea underscores the dangerous circumstances that humans find ourselves in with respect the future of our planet.

8.1 From Biosphere to the Noösphere

Despite the novelty (and controversy) of the Gaia hypothesis, the notion that life plays a critical role in shaping planetary processes was quite explicitly anticipated in the work of several authors in previous decades. These ideas were most conspicuously anticipated in *Biosphere* (1926), a book written by the Russian-born, Ukrainian mineralogist and geologist Vladimir Ivanovich Vernadsky (1863–1945). Although the more familiar use of the term "biosphere" as a collective term for all living matter is a concept attributed to Jean-Baptiste Lamarck (1744–1829), when Vernadsky used this same term, he had in mind several interrelated ideas about how life influences geochemical processes. He wrote, that "…adjusting gradually and slowly, life seized the biosphere and… this process is not over yet" [5]. Life must therefore be considered alongside other forces in shaping global geochemical processes. Organisms in the biosphere exert their influence, not just on spatially restricted and temporally limited scales, but collectively they may regulate biochemical processes (e.g. photosynthesis, decomposition) on large scales [6]. Vernadsky explicitly considered these biospheric influences on vaster geological timescales, and, indeed, on at the spatial scale of the entire planet [6].

In Vernadsky's theorizing on the biosphere, he was himself drawing upon a range of predecessors. In particular, he acknowledged that his thinking was influenced by the work of the Austrian geologist Edward Suess (1831–1914) who coined the term "biosphere," although that concept is not highly developed in Seuss's work [7]. The early Gaian theorists seem to have been unfamiliar with Vernadsky's work when proposing their hypothesis, but later writing emphatically confirmed Vernadsky as an important forerunner of these ideas [7].

In addition to his influential (though for a while occluded) notion of "biosphere," Vernadsky is also associated with the development of the concept of the *noösphere* in the 1920s. The concept of the noösphere emerged in conversations between Vernadsky, the French geologist Pierre Tielhard de Chardin (who coined the term), and philosopher Edouard Le Roy. De Chardin's development of the idea envisioned the noösphere (the zone of human mental influence) as emerging from a progressive developmental sequence starting with lithogenesis (the origins of rocks etc.) through biogenesis (culminating in appearance of the biosphere) and ending with noögenesis (the emergence of the sphere of the mind). De Chardin – a Jesuit priest – sought to reconcile Darwinian evolutionary ideas with notions of a directed progression in earth history (that is, de Chardin's thinking is explicitly teleological). In de Chardin's view, evolutionary processes led to the advent of the human mind – and were ultimately guided by the enigmatic "omega point" ("a superior pole to the world") – which represents an acme of evolutionary development [8]. Vernadsky, a committed empiricist, did not subscribe to the same metaphysical aspirations as de Chardin in developing his conceptual framework. From Vernadsky's perspective, the noösphere was merely an expression of the power of the human mind expressed as a geological force. Human mental facilities are a relative novelty on the planet, and yet, alongside the physical, chemical, and biological forces, mental processes shape the globe. Unlike de Chardin's version of the noösphere, which had both earthbound and transcendental implication, in Vernadsky's version, the noösphere is purely a stage in biogeochemical evolution [9]. A defining moment for the emergence of the noösphere was, for Vernadsky, the point where humankind created energy resources "through the nuclear transmutation of elements" [10]

The precise meaning of *Omega* (and the *Omega Point*, which Pierre Tielhard de Chardin only uses in one footnote) as developed in his book *The Phenomenon of Man* (1959) is somewhat cryptic. De Chardin writes: "By its structure Omega, in its ultimate principle, can only be *a distinct Center radiating at the core of a system of centers*; a grouping in which personalisation of All and personalisation of the elements reach their maximum, simultaneously and without merging, under the influence of a supremely autonomous focus of union." pp. 262–263.

Vernadsky remained relatively optimistic that humans could shape the relevant planetary forces to meet the needs of our own species. From his perspective, the appearance of the noösphere can be thought to inaugurate a new stage of civilization. Despite Vernadsky's optimism about the noösphere, at least one influential American ecologist, Eugene Odum (1913–2002), was concerned that the implications of the noösphere concept. The noösphere seemed to Odum to imply a planet already under the governance of humans. Unlike Vernadsky who was prepared to see this as a beneficent idea, Odum regarded the prospect of human control over the planet as an undesirable thing [11, 12].

8.2 The Geological Ages of Humankind

Stratigraphy is the subdiscipline of geology dealing with sedimentary rock layers (strata). Its basic principle is that the upper layers of a sedimentary sequences were accumulated later in time than the lower ones; thus one can provide an idealized sequence of the geological ages.

The noösphere was by no means the first conceptualization of the human influence on Earth's systems. What is distinctive about it, conceptually, is that both de Chardin's and Vernadsky's version seek to represent the human intellect or consciousness as an emerging biogeochemical force. Although the noösphere has been relegated in the environmental sciences to an intellectual curiosity, it may yet prove its value in understanding our current predicament. Other schemes that account for the humans in a biophysical context do so by announcing the advent of humans expressed in term of stratigraphic history. That is, a geological epoch of humans is said to be upon us. For example, Antonio Stoppani (1824–1891), an Italian geologist and humanist, pronounced the emergence of an *Anthropozoic era*. Stoppani is sometimes referred to as the father of "geoethics" [13, 14] He regarded human beings as "geological agents," and promoted an urgent need for the ethical management of georesources [13]. The human being is, in Stoppani's view, "an agent, perhaps new, certainly still inexperienced of his power, . . . called upon to put the last touches to the work of the ages" [15], (*translation in* [13]). Stoppani's notion of an Anthropozoic era was quoted favorably by George Perkins Marsh in his foundational conservation text *Man and Nature* of 1864 (*The Earth as Modified by Human Action*) [16, 17].

Similar to the Anthropozoic, is the notion of a *Psychozoic era* proposed by Joseph LeConte (1877). LeConte envisioned the emergence of the human being as a geological event [18]. LeConte regarded the era dominated by humans as the fourth geological era (the Paleozoic, Mesozoic, and Cenozoic being the first three). He wrote that one can "regard. . .. the Quaternary [i.e. the most recent of the three periods of the Cenozoic Era] as the critical, revolutionary, or transitional period between [the eras]." By critical, revolutionary, and transitional, LeConte signifies periods where "new and higher organic forms" originate. Eras that follows a critical one are times of "tranquility" [sic] where there is a readjustment to these forms [18]. Importantly, the emergence of the Psychozoic did not coincide with the mere appearance of humans. LeConte wrote: "As fishes existed before the age of fishes, reptiles before the age of reptiles, and mammals before the age of mammals, so man also appeared before the age of man" [18]. The Psychozoic era (the "age of man") thus commences not with the mere existence of humans, but rather with the "completed supremacy" of humans. The term Psychozoic was hardly greeted with enthusiasm the decades after its annunciation. Commenting on the Psychozoic era in the journal *Science*, paleontologist Edward W Berry asked: "It is probably good philosophy to commence earth history with a hypothetical Archeozoic but is it equally good philosophy to terminate earth history with a Psychozoic era?" [19]. His answer was an emphatic no. Though Berry acknowledged "the magnitude and multifarious effects of human activity," he nonetheless concluded that these activities were scarcely of significance on a geologic scale. Objectively, Berry opined that there are other geological events of greater significance than the genesis of humans [19].

The implications of the foregoing are threefold. Firstly, there are important conceptual precedents, developed over the past century, for regarding life as a geological force operating on a global scale. Secondly, the framework of the science of biogeochemistry can extend to include investigations of how humans, as a manifestation of the same evolutionary processes as the rest of the biosphere, can exert pressure on global systems. Finally, these models can be agnostic on whether the human influence on planetary affairs is a positive one or not. Vernadsky, as we have seen, imagined that the noösphere – human mental attributes expressed as a geochemical force – might be beneficial, at least for continuation of the human enterprise.

8.3 From Noösphere to Anthropocene

We have, supposedly, left the Holocene epoch – the geological duration (within the Quaternary Period) that provided the affable climatic conditions of the past 10–12 millennia in which civilization emerged and flourished – and have now entered a new epoch, where humans have become the globally dominant force. The proposed name for the epoch is the Anthropocene. Though the claim that we now live in a new geological epoch remains controversial, a proposal to ratify the term has been submitted to the Subcommission on Quaternary Stratigraphy, whose task is to formalize geological ages, and this is now receiving full consideration [20, 21].

The "Anthropocene" has been proposed as an official unit of the International Chronostratigraphic Chart/Geological Time Scale. Though proposed, the Anthropocene is not currently a formally defined geological unit within the Geological Time Scale. We are still, officially, living within the Meghalayan Age of the Holocene Epoch [1].

The idea of the Anthropocene is clearly related to several of the conceptual frameworks we discussed above: the Noösphere, the Anthropozoic, and the Psychozoic eras. Each of the concepts are developed in the contexts of geology, a consideration of the ages of earth, and the earth sciences broadly understood. All of them, including the more recently proposed Anthropocene, conceptualize a strong influence of humans – considered for this purpose purely as biological entities – on planetary affairs. Although in recent years, scholars of the Anthropocene have been careful to acknowledge and elucidate the precedents of their conceptual models, nonetheless it is important to stress that each of these concepts stakes out a distinctive intellectual domain [9]. For example, though de Chardin's and Vernadsky's versions of the noösphere appear conceptually related – both regard the global influence of the human mind in the context of a sequential (and evolutionary) development of stages in the history of the earth – de Chardin's concept of the noösphere has a spiritual dimension that Vernadsky's does not.

We can, therefore, ask what is the distinctive conceptual domain of the Anthropocene – what does this term imply? The Anthropocene uniquely designates an abrupt, irreversible, and *undesirable* departure from the previous trajectory of earth history. It provides a name for the reality that the Earth has moved out of its current geological epoch, and that human activity was largely responsible for the end of the Holocene [16]. In an early formulation of the idea of the Anthropocene, Paul J. Crutzen, the Dutch, Nobel Prize-winning, atmospheric chemist who popularized the term, wrote that it was appropriate to assign "the term 'Anthropocene' to the present, in many ways human-dominated, geological epoch" [22]. In a paper that makes explicit reference to the antecedent ideas of the Anthropozoic era and the noösphere, Crutzen designated the starting point of the Anthropocene as the latter part of the nineteenth century where global concentrations of carbon dioxide and methane (as analyzed by air trapped in polar ice) were elevated over background levels. Though the Anthropocene is often regarded as synonymous with an era dominated by anthropogenic climate change, it is also more specially understood to designate an era when the actions of humans as a forcing agent leaves a demonstrable geological record [23].

The 'golden spike' marking the Ediacaran GSSP. The Ediacaran Period is a geological period that lasted 94 million years that concluded at beginning of the Cambrian Period 541 Mya. (Ediacara, South Australia). *Source:* Bahudhara / Wikimedia Commons / CC BY-SA 3.0.

A Global Standard Stratigraphic Age (GSSA), is a chronological reference point that can be used to define boundaries between different geological durations.

Viewed in this way, we can ask: what legacy does recent human activity leave in the geophysical record? Or to paraphrase Edward W. Berry, "is it good philosophy to terminate earth history with an Anthropocene era?" In terms consistent with the designation of other geological ages we should also be able to identify global stratigraphic markers of the event (technically the Global Strato-type Section and Point [GSSP]) that indicates a change in the Earth system significant enough to warrant naming a new epoch. The transition between one geological age and another is marked with a (typically figurative) "golden spike" [24].

The GSSP can be a distinctive fossil or physical event (a geomagnetic reversal, or a globally distributed change in stable isotope values, for example) that is identifiable consistently in the rock record, and marks the lower boundary of a geological stage [25]. Typically, multiple criteria are identified; these are ones that can be found in locations across the world. What, then, is the golden spike marking the Anthropocene?

8.4 A "Golden Spike" for the Anthropocene

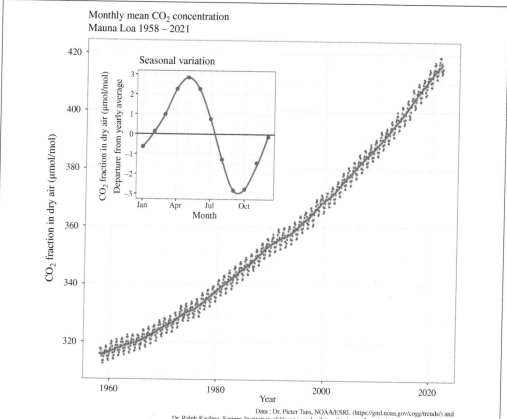

Monthly mean CO_2 concentration
Mauna Loa 1958 – 2021

Data : Dr. Pieter Tans, NOAA/ESRL (https://gml.noaa.gov/ccgg/trends/) and
Dr. Ralph Keeling, Scripps Institution of Oceanography (https://scrippsco2.ucsd.edu/). Accessed 2021-12-16
https://w.wiki/4ZWn

Atmospheric CO_2 recorded at Mauna Loa Observatory. This figure shows the history of atmospheric carbon dioxide concentrations as directly measured at Mauna Loa, Hawaii, since 1958. This curve is known as the Keeling curve, and is an essential piece of evidence of the human-made increases in greenhouse gases that are believed to be the cause of global warming. Changes in CO_2 levels are often used to signify the emergence of the Anthropocene.

The defining event that is often used to date the genesis of the Anthropocene, and the one used by Crutzen and many other scientists, is the detection of elevated concentrations of carbon dioxide (CO_2) in the atmosphere compared to pre-industrial times. CO_2 concentrations above the pre-industrial levels have been extensively monitored since Charles Keeling's initial observation at Hawaii's Mauna Loa observatory [26]. Keeling's 1976 summary of these early investigations, which I will reproduce in full, can be regarded as one of most significant observations of twentieth century environmental science:

> The concentration of atmospheric carbon dioxide at Mauna Loa Observatory, Hawaii is reported for eight years (1964–1971) of a long-term program to document the effects of the combustion of coal, petroleum, and natural gas on the distribution of CO_2, in the atmosphere. The new data, when combined with earlier data, indicate that the annual average CO_2, concentration rose 3.4 % between 1959 and 1971. The rate of rise, however, has not been steady. In the mid-1960s it declined. Recently it has accelerated. Similar changes in rate have been observed at the South Pole and are evidently a global phenomenon [26].

Since Keeling initiated his foundational work, a system of reliable and precise in situ recording of CO_2 (and other greenhouse gasses) has been developed and deployed at a variety of sites (one significant system is the National Oceanic and Aeronautic Administration's (NOAA) Earth System Research Laboratory's (ESRL) Global Greenhouse Gas Reference Network [27]). The relationship between elevated atmospheric greenhouse gas concentrations and temperature – a foundational observation for atmospheric and climatological sciences – has long been established [28]. This well characterized relationship between the radiative forcing of greenhouse gases (especially CO_2) and global temperature, is the basis for recent climatological research [29]. The implication of these observations will be considered in more detail in a later chapter, but we should pause here to consider if the elevation in CO_2 will be significant enough to leave a geological record? That is, how might the elevated CO_2 of the present era be detected by a historic geologist inspecting from some imagined future?

It is not easy to answer this question. If CO_2 is elevated in the atmosphere for a relatively short period, judged on the temporal scales typically resolved in geological studies, this will be too evanescent to be detected [30]. However, if CO_2 concentrations are elevated for a considerably longer period (centuries), then a record of this elevated carbon will certainly will be recorded in the stratigraphic sequence.

Generally, the history of ancient atmospheric CO_2 (recorded as the partial pressure of CO_2 [pCO_2]), is evaluated by a range of proxy measures [31]. Such measures include the estimation of stomatal density in fossil leaves and the use of stomatal indices (that is, calculating the percentage stomata out of the total number of epidermal cells plus stomata on those leaves). Stomata are the tiny pores found embedded in the epidermis of leaves, stems, and other organs that serve to regulate rates of gas exchange. Since the density and stomatal index of plants has decreased over the past 200 years in response to elevated atmospheric carbon this may be detectable in plant remains that will likely endure into the future [32]. In addition to future fossil analysis, the analysis of stable carbon-isotope compositions of carbonates in ancient soils (palaeosols) can reveal the atmospheric carbon concentration when these are formed [33]. Finally, the partial pressure of CO_2 can be detected in the boron isotopic compositions of foraminifera, this is because boron isotopic composition of foraminiferal tests (shells) depends on the pH of sea water, which reflects the atmospheric conditions (including its CO_2 content that existed at the time that the foraminifera lived [31, 34]). All of this is to say that a variety of approaches exist that allow for the elucidation of atmospheric conditions at the time that ancient sediments accumulated, and that collectively these lay down a record that can be interpreted in future times.

What other stratigraphically relevant events might serve to demarcate the lower boundary of the Anthropocene? The following markers have been suggested: the accumulation of plastic waste, or more generally, the proliferation of new human-made materials that might leave a record in the form of "technofossils" [35, 36]; the depositions of artificial radionuclides, especially from the detonation and testing of atomic bombs (in the form of ^{137}Cs (cesium) and $^{239+240}$Pu (plutonium) radiogenic fallout from nuclear weapons testing [37, 38]); indirect markers from an alteration in the nitrogen cycle (stable isotope ratios (^{15}N : ^{14}N), which can reveal "a coherent signal" of distinct anthropogenic sources of nitrogen to ecosystems, beginning at least from the early twentieth century [39]; contamination of trace metals (aluminum, silver, chromium copper, nickel, lead, zinc, etc.) [36]; a pattern to the redistribution of a wide variety of material that is distinctive (though natural processes redistribute material, human alteration of flows can be move several times the amount of material transported by the world's rivers [25]); and finally the contemporary loss of species and redistribution of biota globally is likely to leave a stratigraphically detectable signature [40–42].

8.5 The Anthropocene Defined

Ample evidence, as we have seen, can be marshaled to support the designation of the new Anthropocene epoch [25, 43]. That being stated, the proposal for formally recognizing the Anthropocene is unusual in a number of respects. Typically, discussions about the formal designation of geological strata are conducted among historical geologists and concern *ancient* geological events. The events that illustrate the unfolding of the Anthropocene, however, include phenomena both of a sedimentary nature, as well as certain ecological occurrences, neither of which would ordinarily be considered by traditional stratigraphic methods.

Discourse concerning the Anthropocene and data relevant to determining this recent and supposedly irreversibly transition in Earth history engages multiple disciplines: ecology, biodiversity studies, climatology, biogeochemistry as well as the social scientific and humanistic disciplines [44]. Thus, the application of methodology usually availed of by the International Union of Geological Sciences (IUGS) in determining geochronological matters on a global scale, may not capture the peculiar nature of this recent and ongoing change in the earth system [21, 45].

Efforts to find a distinct geological marker – a "golden spike" – for the Anthropocene are further complicated by the fact that events associated with the Anthropocene occur at different rates in different places – that is, they are time-transgressive (or diachronous, to use geological terminology) in onset and development [46]. The Anthropocene is unfolding on archeological and recent historical timescales and are occurring, spatially, at different rates [47]. Early efforts to assign the emergence of the Anthropocene to say, the industrial period with its elevation in greenhouse gas emissions [22], or to isotopic signatures of atomic fallout [37], may be overly simplistic for a time-transgressive phenomena. Therefore, any search for a simple and widespread boundary for the Anthropocene may be misguided.

A more informal definition of the Anthropocene would facilitate flexibility when assigning a beginning date for the epoch [48]. Flexibility in assigning a beginning to the Anthropocene might also focus more research on the geophysical consequences of events in a deeper historical past – especially those that might signify an early beginning for the transformation of the planet by humans [48]. Such events might include the putative extinction of megafauna through human predation [49], the domestication of plants and animals [50] – both of which occurred thousands, if not tens of thousands, of years ago – as well as the accompanying transformations of landscapes that made way for agricultural economies. These elements encoded in an informal concept Anthropocene bring attention to the radical transformation of the planetary systems that date almost to the beginning of the Holocene. We humans may have influenced planetary affairs long before we had previously imagined.

8.6 Coda

Whether or not the Anthropocene is recognized as a formal stratigraphic period may not be as significant, ultimately, as the recognition that just as life, in general, transformed the geophysical nature of the planet, the activities of just one species – a relatively recently evolved primate – has, in the very recent past, deepened the transformation of the planet. Humans have done so in a manner that is plausibly fateful not only to our own flourishing but to the well-being of most other living things on Planet Earth.

Should the emergence of the Anthropocene be regarded as simply an evolutionary development in the story of life on the planet, or should it, rather, be considered as a potentially dangerous approaching by humans to the limits of the planet? It may, of course, represent both of these states of affairs.

References

1 Subcommission on Quaternary Stratigraphy (SQS)/International Commission on Stratigraphy (ICS)/ International Union of Geological Sciences (IUGS) (2019). Working Group on the 'Anthropocene'. http://quaternary.stratigraphy.org/working-groups/anthropocene (accessed 16 November 2022).

2 Lovelock, J.E. (1979). *Gaia: A New Look at Life on Earth*. Oxford Paperbacks.

3 Margulis, L. and Lovelock, J.E. (1974). Biological modulation of the Earth's atmosphere. *Icarus* 21 (4): 471–489.

4 Volk, T. (2012). *Gaia's Body: Toward a Physiology of Earth*. Springer Science & Business Media.

5 Vernadsky, V.I. (1998). *The Biosphere*. Springer Science & Business Media.

6 Ghilarov, A.M. (1995). Vernadsky's biosphere concept: an historical perspective. *The Quarterly Review of Biology* 70 (2): 193–203.

7 Polunin, N. and Grinevald, J. (1988). Vernadsky and biospheral ecology. *Environmental Conservation* 15 (2): 117–122.

8 De Chardin, P.T. (2018). *The Phenomenon of Man*. Lulu Press, Inc.

9 Hamilton, C. and Grinevald, J. (2015). Was the Anthropocene anticipated? *The Anthropocene Review* 2 (1): 59–72.

10 Kautzleben, H. and Müller, A. (2014). Vladimir Ivanovich Vernadsky (1863–1945) — from mineral to noosphere. *Journal of Geochemical Exploration* 147: 4–10.

11 Turner, D.P. (2005). Thinking at the global scale. *Global Ecology and Biogeography* 14 (6): 505–508.

12 Odum, E.P. (1959). *Fundamentals of Ecology*. W.B. Saunders.

13 Lucchesi, S. (2017). Geosciences at the service of society: the path traced by Antonio Stoppani. *Annals of Geophysics* 60: 7.

14 Peppoloni, S. and Di Capua, G. (2017). Geoethics: ethical, social and cultural implications in geosciences. *Annales de Geophysique* https://doi.org/10.4401/ag-7473.

15 Stoppani, A. (1873). *Corso di Geologia*, vol. 3. Milano: G Bernardoni e G Brigola.

16 Steffen, W., Grinevald, J., Crutzen, P., and McNeill, J. (2011). The Anthropocene: conceptual and historical perspectives. *P.hilosophical Transactions of the Royal Society A: Mathematical, Physical and Engineering Sciences* 369 (1938): 842–867.

17 Marsh, G.P. (1874). *The Earth as Modified by Human Action*. Ayer Company Pub.

18 Leconte, J. (1877). On critical periods in the history of the earth, and their relation to evolution; on the Quarternary as such a period. *The American Naturalist* 11 (9): 540–557.

19 Berry, E.W. (1926). The term psychozoic. *Science* 64 (1644): 16.

20 Gibbard, P.L. and Walker, M. (2014). The term 'Anthropocene' in the context of formal geological classification. *Geological Society, London, Special Publications* 395 (1): 29–37.

21 Ruddiman, W.F. (2018). Three flaws in defining a formal 'Anthropocene'. *Progress in Physical Geography: Earth and Environment* 42 (4): 451–461.

22 Crutzen, P.J. (2002). Geology of mankind. *Nature* 415 (6867): 23.

23 Ruddiman, W.F. (2013). The Anthropocene. In: *Annual Review of Earth and Planetary Sciences*, vol. 41. Palo Alto: Annual Reviews (ed. R. Jeanloz), 45–68.

24 Lewis, S.L. and Maslin, M.A. (2015). Defining the anthropocene. *Nature* 519 (7542): 171–180.

25 Elias, S. (2017). Finding a "Golden Spike" to mark the Anthropocene. *Encyclopedia of the Anthropocene* 1: 19.

26 Keeling, C.D., Bacastow, R.B., Bainbridge, A.E. et al. (1976). Atmospheric carbon dioxide variations at Mauna Loa observatory, Hawaii. *Tellus* 28 (6): 538–551.

27 Andrews, A., Kofler, J., Trudeau, M. et al. (2014). CO_2, CO, and CH_4 measurements from tall towers in the NOAA earth system research Laboratory's global greenhouse gas reference network: instrumentation, uncertainty analysis, and recommendations for future high-accuracy greenhouse gas monitoring efforts. *Atmospheric Measurement Techniques* 7 (2): 647.

28 Emanuel, K. (2018). *What we Know about Climate Change*. MIT Press.

29 Rasmussen, S.O., Andersen, K.K., Svensson, A.M. et al. (2006). A new Greenland ice core chronology for the last glacial termination. *Journal of Geophysical Research* 111 (D6).

30 Brannen P. (2019) The Anthropocene is a joke. On geological timescales, human civilization is an event, not an epoch. *The Atlantic* (13 August).

31 Zhang, Y.G., Pagani, M., Liu, Z. et al. (2013). A 40-million-year history of atmospheric CO_2. *Philosophical Transactions of the Royal Society A: Mathematical, Physical and Engineering Sciences* 371 (2001): 20130096.

32 McElwain, J. (1995). Stomatal density and index of fossil plants track atmospheric carbon dioxide in the Palaeozoic. *Annals of Botany* 76 (4): 389–395.

33 Mora, C.I., Driese, S.G., and Seager, P.G. (1991). Carbon dioxide in the Paleozoic atmosphere: evidence from carbon-isotope compositions of pedogenic carbonate. *Geology* 19 (10): 1017–1020.

34 Spivack, A.J., You, C.-F., C.-F., and Smith, H.J. (1993). Foraminiferal boron isotope ratios as a proxy for surface ocean pH over the past 21 Myr. *Nature* 363 (6425): 149–151.

35 Zalasiewicz, J., Waters, C.N., Ivar Do Sul, J.A. et al. (2016). The geological cycle of plastics and their use as a stratigraphic indicator of the Anthropocene. *Anthropocene* 13: 4–17.

36 Zalasiewicz, J., Williams, M., Waters, C.N. et al. (2014). The technofossil record of humans. *The Anthropocene Review* 1 (1): 34–43.

37 Waters, C.N., Syvitski, J.P.M., Gałuszka, A. et al. (2015). Can nuclear weapons fallout mark the beginning of the Anthropocene epoch? *Bulletin of the Atomic Scientists* 71 (3): 46–57.

38 Dean, J.R., Leng, M.J., and Mackay, A.W. (2014). Is there an isotopic signature of the Anthropocene? *The Anthropocene Review* 1 (3): 276–287.

39 Holtgrieve, G.W., Schindler, D.E., Hobbs, W.O. et al. (2011). A coherent signature of anthropogenic nitrogen deposition to remote watersheds of the northern hemisphere. *Science* 334 (6062): 1545–1548.

40 Zalasiewicz, J., Williams, M., Smith, A. et al. (2008). Are we now living in the Anthropocene? *GSA Today* 18 (2): 4.

41 Barnosky, A.D., Hadly, E.A., Dirzo, R. et al. (2014). Translating science for decision makers to help navigate the Anthropocene. *The Anthropocene Review* 1 (2): 160–170.

42 Wilkinson, I.P., Poirier, C., Head, M.J. et al. (2014). Microbiotic signatures of the Anthropocene in marginal marine and freshwater palaeoenvironments. *Geological Society, London, Special Publications* 395 (1): 185–219.

43 Zalasiewicz, J. and Waters, C. (2018). Arguments for a formal global boundary Stratotype section and point for the Anthropocene. *Encyclopedia of The Anthropocene* 1: 29–34.

44 Inkpen, S.A. and Desroches, C.T. (2019). Revamping the image of science for the Anthropocene. *Philosophy, Theory, and Practice in Biology* 11: 20200929.

45 Zalasiewicz, J., Waters, C.N., Head, M.J. et al. (2019). A formal Anthropocene is compatible with but distinct from its diachronous anthropogenic counterparts: a response to W.F. Ruddiman's 'three flaws in defining a formal Anthropocene'. *Progress in Physical Geography: Earth and Environment* 43 (3): 319–333.

46 Edgeworth, M., Deb Richter, D., Waters, C. et al. (2015). Diachronous beginnings of the Anthropocene: the lower bounding surface of anthropogenic deposits. *The Anthropocene Review* 2 (1): 33–58.

47 Edgeworth, M., Ellis, E.C., Gibbard, P. et al. (2019). The chronostratigraphic method is unsuitable for determining the start of the Anthropocene. *Progress in Physical Geography: Earth and Environment* 43 (3): 334–344.

48 Rull, V. (2016). The humanized earth system (HES). *Holocene* 26 (9): 1513–1516.

49 Doughty, C.E., Wolf, A., and Field, C.B. (2010). Biophysical feedbacks between the Pleistocene megafauna extinction and climate: the first human-induced global warming? *Geophysical Research Letters* 37: L15703. https://doi.org/10.1029/2010GL043985.

50 Smith, B.D. and Zeder, M.A. (2013). The onset of the Anthropocene. *Anthropocene* 4: 8–13.

Section Four

The Concept of Limits

9

The Anthropocene and the Concept of Limits

To live on a materially finite planet is to live within limits [1]. This is both a thermodynamic necessity and a practical truism. Living within limits is a statement that is both necessary and true in the sense that the conduct of human affairs must be constrained, as all things are, by the fundamental laws of physics. These limiting necessities govern our use of the energy fixed by plants, as well as in our exploitation of both the stored energy and the material reserves of the Earth.

Let us briefly examine how physical laws dictate the necessity of living within limits. The *first law of thermodynamics* states that the energy of the universe remains constant, energy being neither created nor destroyed. As we exploit energy to do work, heat is either stored or dissipated. By ecological definition, a resource is depleted when used – that is, consumption by one organism leaves less for another, unless and until, that is, the resource is replenished by the naturally occurring processes. The fixing of energy by photosynthesis, for example, is limited by many factors, including the finite rates of key nutrient cycles. These rates of nutrient cycling are, in turn, limited and energized by flows of solar radiation. Plant life sustains its consumers, therefore, only up to the limits of its continued productivity. Excess consumption over production can profoundly affect the structure of ecological communities. Conforming to the *second law of thermodynamics*, humans, like any other creatures, must, in their appropriation of resources, create a measure of disorder in their environment. This disordering may alter the behavior of key biogeochemical processes. *Feeding on order* to *maintain order*, as we have previously seen, is the essence of the living being: this principle is at the core of Erwin Schrödinger's answer to the question, what is life? [2]. All things being equal, the greater the scale of human ecological activity, the greater is the risk from inescapable environmental disordering. In satisfying our needs (and wants) by exploiting scarce resources – that is, in being *economic* creatures – we must adapt to the finiteness of resources (substituting new resources, when this is needed), and, in doing so, we rely upon the continued productivity of nature. We must also remediate the disorder (entropy) produced by the human exploitation of resources, before risking a rupturing in the capacity of natural capital to replenish, albeit at a finite rate, the resources upon which all living things rely. Entropy cannot be halted, of course, but disrepair can be managed.

A Primer on Human Impacts on the Environment: The Conceptual Approach, First Edition. Liam Heneghan.
© 2023 John Wiley & Sons Ltd. Published 2023 by John Wiley & Sons Ltd.

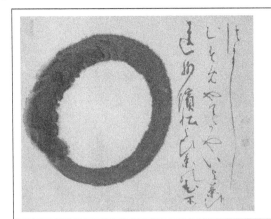

The circle in Zen art is known as "enso." It is often called the "circle of enlightenment." Enzo is a theme for contemplation; such drawings are often accompanied by a short spiritual text. The circle is also a frequent motif in environmental thought, representing the world as habitat, but also indicating the finiteness of the Earth. *Source:* Hakuin Ekaku/Wikimedia Commons / CC BY-SA 4.0.

There is no universally accepted definition of *economics*. One famous definition, provided by the British economist Lionel Robbins, is that economics is "the science which studies human behavior as a relationship between ends and scarce means which have alternative uses." The definition is helpful because it underscores the fact that both economics and ecology are committed to the study of scarcity.

In this and the next chapter we ask if human economic and social enterprises can continue to expand along their dominant lines – capitalism, socialism, or indeed any other highly consumptive economic system – without risking catastrophic consequences for human health and well-being, and for the fate of many other species. To phrase it as a number of environmental modelers and ecological economists have, we ask if we, the humans, live in an *empty world* (where economic growth – business-as-usual – can continue) or a *full world* where we have met, or even breached, planetary limits [3, 4]. By "limit" what is meant here is a boundary point that cannot be breached without consequence. Limits on the expansion of human economic development, traditionally understood, may therefore occur through the exhaustion of environmental resources, *or* through the depletion of an environmental sink. The latter source of limits on economic growth and development has been less often appreciated.

Natural capital represents the reserves of natural ecosystems that that can sustainably produce ecosystem goods or services.

The notion of limits is central to both ecological and economic thought: it is the unifying thread connecting the two disciplines. In what follows here we excavate, briefly, how the concept of limits has been central to an understanding of the structure of ecological communities and to understanding the population dynamics both of wild creatures, and of the human being. In the next chapter, we sift through the evidence asserting that we have, indeed, breached some significant limits.

9.1   The Meaning of Limits: Malthus, Darwin, and Contemporary Ecology

Limits can be understood in a variety of ways. Thought of in ecological terms, an important meaning of "limit" – viz. the regulatory control imposed by the finiteness of natural resources – has been employed in *demographic analysis* as a means by which fluctuations in the population of any species might be understood, and in *ecological community analysis* to understand the mechanisms that structure multispecies assemblages of organisms. Investigating the degree to which population growth is limited by resources (alongside other ecological factors) has been a cornerstone of both population and community ecological research [5]. For instance, in a foundational 1960 research paper, "Community Structure, Population Control, and Competition," ecologists Nelson G. Hairston, Frederick E. Smith and Lawrence B. Slobodkin wrote,

> There thus exists either direct proof or a great preponderance of factual evidence that in terrestrial communities decomposers, producers, and predators, as whole trophic levels, are *resource-limited* in the classical density-dependent fashion. Each of these three can and does expand toward the *limit* of the appropriate resource [emphasis mine] [5].

> **Demography** is the statistical study of populations and examining the changing size or composition of populations over time.

As the population of any species exhausts a given resource, the per capita population growth rate decreases. This can be understood fairly intuitively, after all a population grows only in as much as the resources of its environment can support this growth. Resource exhaustion imposes a regulatory feedback on population size. This mechanism of density-dependent resource limitation – a *negative feedback process* – arises naturally and, by this means, resource availability can affect survival and reproductive rates, and hence the growth of a population. The concept of resource limitation as a regulatory force remains important in contemporary ecological theorizing; yet, these ideas have a long history. Conjectures about how the growth of populations may be limited by resources dates back right into the earliest history of both ecology and economics as academic disciplines.

SOME HANDY DEMOGRAPHIC TERMINOLOGY

Density-dependent population growth is when the per capita population growth rate changes as population density changes. As the death rates rise or the birth rates fall, this is manifested in diminished population growth.

   Per capita *population growth rate* is the rate at which a population size changes expressed per an individual in the population.

   Negative feedback: in ecology negative feedback is a response that inhibits or modifies inputs into that system – when a population increases, for example, this can create conditions that negatively affect birth rates, or increase death rates. **Positive feedback**, on the other hand, occurs when outputs amplify inputs into that system (and can cause a runaway change).

The term ecology was coined by the biologist Ernst Haeckel (1834–1919) in 1866 [6]; Darwin, therefore, does not use the term "ecology." Nonetheless, *On the Origin of Species* (1859) can undoubtedly be thought of as an early ecological text. This is because both Darwin and Alfred Russel Wallace, the cooriginators of the concept of natural selection, perceived that the transformation of species emerges from ecological interactions among individual organisms that occur in local populations during the "struggle for existence." The thinking of limits is omnipresent in Darwin's treatise. For example, in Chapter III, *The Struggle for Existence* – the ecological crux of Darwin's argument – he writes:

> The amount of food for each species, of course, gives the extreme *limit* to which each can increase; but very frequently it is not the obtaining food, but the serving as prey to other animals, which determines the average number of a species [emphasis mine] [7].

Darwin was quite emphatic about the role that resource limitation plays in driving changes in nature. "The amount of food for each species of course gives the extreme limit to which each can increase. . ." Darwin observed. Though resources are undoubtedly important, it is clear that Darwin recognized that other factors may regulate populations. This is what is meant in the quote above ". . .but very frequently it is not the obtaining food, but the serving as prey to other animals, which determines the average number of a species." [7]

In speculating on the theoretical significance of limited access to resources, Darwin is echoing the ideas of an important predecessor: the economist and political theorist Thomas Malthus (1766–1834). Though his influence is quite often regarded as a baleful one, Malthus is best known for his writing on the importance of resources as setting limits on the growth of the *human* population. References to "limit" (and variations thereof) occur more than 20 times in Malthus's most renowned text, *Essay on the Principle of Population* (published anonymously in 1798) [8]. Though Malthus cannot be considered an ecologist, this work (and his meditations on the limits of human population growth) affected the development of evolutionary studies through his influence on both Charles Darwin's and Alfred Russel Wallace's notion of natural selection [9]. In setting forth his understanding of how human populations are regulated, Malthus wrote:

> Must it not then be acknowledged by an attentive examiner of the histories of mankind, that in every age and in every state in which man has existed, or does now exist. . .. that the increase of population is necessarily *limited* by the means of subsistence. . . that population does invariably increase when the means of subsistence increase. . . [and]. . . that the superior power of population is repressed, and the actual population kept equal to the means of subsistence, by misery and vice? [emphasis authors] [8].

It remains a matter of debate, among environmental historians, the degree to which Malthus in writing about human populations provided a *direct* contribution to Darwin's formulation of natural selection. In describing the struggle for existence in the context of competition among individuals for access to limited "subsistence" Darwin referenced Malthus, though he may simply have been using Malthus to confirm an opinion that he had already developed about the significance for organic development of a "struggle for existence" [9].

The **exponential growth** of a population is where future growth of that population depends upon its size at a given time; mathematically, such growth is given by the equation $y = ae^{bt}$, where "a" and "b" are constants and $a > 0$. "e" is the exponential constant (its value is approximately 2.718).

For all of this, Darwin, in a sense, is Malthus writ large and applied to the whole of the animal and vegetable kingdoms, and therefore not merely to human populations. In *On the Origin of Species* Darwin draws out the full implications of the propensity of populations to grow geometrically (or what today we would call exponential growth) in contrast to the potentially linear patterns of growth (arithmetic growth, Malthus calls it) of the subsistence upon which living beings rely. Ultimately, competition for limited resources – what Darwin terms the struggle for existence – means that variation beneficial to an individual "will tend to the preservation of that individual, and will generally be inherited by its offspring" [7]. Natural selection is driven by a Malthusian engine, and, fundamentally, Darwin's theory, as we have argued here, is an ecological one [10].

Undoubtedly, the significance of Malthus's discussion of "limits" in relation to human populations was amplified by Darwin (and Alfred Russel Wallace [11]) and was thus transformed into a far ranging hypothesis governing all of nature. In as much as humans are products of the same natural forces and processes as the rest of nature, it should be no surprise that Malthus' speculation about the nature of population growth should apply with equal force to the "race of plants and the race of animals" [8]. Nonetheless, it is Malthus' emphasis on the factors limiting to human population growth and the prospects for "unconceived improvement" in the human condition that is immediately relevant to the issues considered in this and the next chapter, namely, the question concerning the degree to which humans are constrained within planetary limits. It is, therefore, Malthus and not Darwin that is most often considered in discussions about the limits to growth.

The work of Malthus has remained controversial in the more than two centuries since it was written – Karl Marx, for example, regarded Malthus work as "nothing more than a schoolboyish, superficial plagiary. . ." A main complaint directed against Malthus, both in his own time and in present debates over population growth, concerns his conclusion that "poor laws" (these were systems of relief used to alleviate poverty, and codified into law in England and elsewhere) are always self-defeating. This is because, according to Malthus, when conditions of poverty are eased, and societal limits are lifted, the resulting access to additional resources induces that "part of society that cannot in general be considered as the most valuable part" to have larger families. Just as controversial today as it was in eighteenth and nineteenth centuries, are questions concerning the implications of providing famine relief and resource assistance to regions where population growth rates remain high. Does such relief exacerbate population growth, as contemporary disciples of Malthus ("neo-Malthusians" as these followers are sometimes called) claim? Alternatively, might relief lead to an increase in the wealth of nations and thereby lead to lower population growth, as has been shown to be the case when resource availability leads to wealth creation [12]. Malthus' central scientific central thesis, that *the power of population is indefinitely greater than the power in the earth to produce subsistence*, is contentious for scientific reasons, as well as for reasons of questionable social policy – since even early ecologists recognized that the regulation of population size in other species is a complicated matter [13]. Nevertheless, the essential claim of an incommensurability between increasing demands for subsistence through "geometric" growth of population with the, at best, arithmetically (that is, linearly) growing supply of subsistence, has remained stimulating for environmental modeling conducted over the past half century [14].

Malthus uses the term **subsistence** extensively throughout *An Essay on the Principle of Population* (1798) without providing a technical definition. It is clear that Malthus uses the term in a generic sense and is ultimately referring to the means of supporting the life of a person, that is, water and food and the processes that provide these.

9.2   Limits of the Perfectibility of the Human Being: Energy and Cultural Development

In his book *Enquiry Concerning Political Justice and its Influence on Morals and Happiness* (1793) the English political philosopher and writer William Godwin's opined on the "perfectibility" of the human condition [15]. Conjectures on the possibly infinite improvement of the human lot – especially in response to Godwin's optimism on the matter – has inspired much thinking. Until the present day, many writers are convinced that the human story is one of infinite progress. Such faith in progress is expressed in one of Godwin's fundamental propositions: "Man is perfectible, or in other words susceptible to perpetual improvements." In addressing the question of the human population and its needs, Godwin seemed inclined to the view that the population would "tend to its own level, and. . .proceed in the most auspicious way, when least interfered with by the mode of regulation." That is, consistent with his politically anarchistic viewpoint, Godwin assumed that processes pertinent to population growth (marriage etc.) should be freed from intervention by political institutions. Aside from exceptional circumstances, the population of a region might undergo very little change over time. Godwin goes on to assure his readers that *even if* populations were to increase considerably, three quarters of the globe remained uninhabitable (at the time of his writing) and thus one should not fret about the possibility of growth. The earth, he insisted, will continue to be "found sufficient for the support of its inhabitants." The human enterprise might continue to grow and improve without discovering itself limited in any unsurpassable way.

Such speculations by Godwin seems to have provided an immediate goad for the development of Thomas Malthus' more pessimistic views on that topic – the improvement of human condition will always be undermined by a burgeoning of the population [15, 16]. Malthus sought to undermine this optimistic assessment by means of the simple model we introduced above and argued for the inevitability of a clash between population growth (geometric) and the ability to furnish subsistence (which grows linearly, at best).

In his *Essay on Population* (1898) Malthus wrote: "Necessity, that imperious all pervading law of nature, restrains [the animal and vegetable kingdoms], within the prescribed bounds. The race of plants and the race of animals shrink under this great restrictive law. And the race of man cannot, by any efforts of reason, escape from it. Among plants and animals its effects are waste of seed, sickness, and premature death. Among mankind, misery and vice."

By these terms **Misery and Vice** Malthus means starvation, disease and death on the one hand and the variety of social ills, including infanticide and "violations of the marriage bed," and on the other.

Malthus' model – the power of which is in its great simplicity – attempted to undermine Godwin's rosy views on perfectibility in the social development of humankind. It suggests a mechanism that can limit population, positing that as limits are approached humans fall into both "misery and vice" [see the panel above for an explanation of these terms].

Despite the grimness of his outlook, Malthus does not, however, provide any specific insight into how the relationship between population growth and subsistence might affect human social development. A more complex model than that of Malthus was developed by the anthropologist Leslie A. White (1900–1975). This model shows the conditions under which the ongoing use of resources can result in an improvement in the human condition and, in so doing, provides insights into the mechanisms of cultural development. White was considered an iconoclastic anthropologist; his approach to the discipline emphasized a revival of nineteenth-century evolutionary thinking in anthropology [17]. In an influential paper entitled, "Energy and the evolution of culture," published in 1943, White conjectured that the development of human culture depends upon "the material, mechanical means by which man articulates himself with the earth" [18]. By this curious phrase – "articulating with the earth" – White meant to direct attention to how the human organism in its habitat expends energy to produce a range of artifacts that serve human needs. Energy – defined as the ability to do work (including the transformation of habitat) – is White's universal currency. White relates the following terms in his conceptual model: energy use per capita (E), the technological means of expending that energy (T), and the magnitude of cultural product per unit time (P) as:

$$ExT = P$$

Humans have a limited amount of bodily energy available to them – the strength of muscle and sinew (sometimes called somatic energy) – thus, the only way that we can significantly extend our ability to increase the rate of production of cultural outputs (beyond getting stronger, or by employing or coercing more labor) is by recourse to energetic resources *beyond* the physical capabilities of humans. For example, humans can enhance the availability of energy to do work by employing animal labor, or by discovering new sources of energy, fossil fuel, nuclear energy, and so on. Additionally, innovation and increased efficiency in the technological means by which these energy sources are harnessed can also increase cultural output.

Leslie White synthesizes these insights in his law of cultural evolution:

> Culture develops when the amount of energy harnessed by man per capita per year is increased; or as the efficiency of the technological means of putting this energy to work is increased; or, as both factors are simultaneously increased [18].

The significance of these observations are threefold.

Firstly, as energy use increases, or as the technological system becomes more efficient, the ability for humans to effectively transform environmental resources into cultural product increases. That is to say, the relationship between the population and its resources is mediated by the complexities of technological innovation, and by their ability to realize cultural ambitions through a capacity to do work, by means of access to various sources of energy. Culture is thus, in White's view, a thermodynamic system that relates energy, tools, and product [19]. In the classic presentation of the model, White has little to say about the potential exhaustibility of resources – that is the limits of subsistence – but he does reflect that an increase in cultural production over time has implications for the human habitat.

Secondly, there is a *limit to how much cultural product can develop* in the context of any given energy source and technological system. For example, the finite limits of human bodily power were overcome rather late in the history of humanity by harnessing the power of domesticated animals (cattle, horses, and so on) [20]. In turn, no matter how much animals are improved through breeding, there is a limit – one that is ultimately imposed by nature – to how much work can be performed. With this is mind, White claimed that after the domestication of plants and animals, and subsequent to initial refinements of that systems, there was little improvement in the "agricultural arts" from 2000 BCE to 1800 CE. When the limits to energy extraction and the refinement of tools within a social system had been realized, little cultural expansion and development was expected or possible. When this limit is reached, cultural evolution ceases until there is a revolution is the energy/technological system.

Finally, the level of innovation that flourished after the onset of the industrial revolution was motivated both by the exploitation of new energy sources *and* the development of new technologies that could be harnessed to the availability of fossil fuels. The advent of nuclear energy, and the newer forms of renewable energy that have recently become widely available, have – according to the terms of White's model – implications, in turn, for cultural progress, and for the types of products made available for work.

White's model accounts for the increases in population apparent as culture develops. As the model implies, population growth is facilitated by the increased availability of subsistent associated with changes in the energetic and technological systems. White quotes, approvingly, the observation by the Victorian era anthropologist Edward Burnett Tylor (1832–1917) that the best means of fully understanding the significance of coal to the industrial era is to recognize that of any three English people in late nineteenth century "one at least may be reckoned to live by coal." That is, the availability of a richer energy source, has implication for population growth. Thus, as in Malthus' model, so also in White's model: population is limited by subsistence. In the latter model, however, the control of population by subsistence is determined by the limits of efficiency of different technological systems. Culture and population does not proceed linearly; rather it proceeds by revolutionary steps. Each revolution is followed by a subsequent realization of the limits to the efficiency of extraction and production.

It might seem on initial inspection that Leslie White shared an optimistic vision of progress (the perfectibility of humans) with the likes of Godwin (and anthropologists like Edward Taylor). Certainly his model suggests a progressive development of culture, and yet while White illustrates how new technological development can allow for greater cultural production, he acknowledged that such developments can be to the ultimate detriment of humans. Humans under the new modes of agriculture (after the domestication of plants and animals) were often reduced, White wrote, "to slavery or serfdom" [21, 22]. This serves, therefore, as a warning that cultural development may proceed from revolutionary changes in technological systems based upon accelerated extraction of resources, but that such development can be associated with environmental impacts and, through modifications of social conditions, a variety of human costs.

At the core of Leslie White's energy model of cultural development is the idea that with each major revolution in technological systems and transformations in energy sources, fresh opportunities are presented for growth in cultural production [18]. At the beginning of each new era – the domestication (agricultural) revolution, the industrial revolution, and so on – novel tools become available that are refined and made more efficient over time. But ultimately, regardless of the technological systems involved, a limit is inevitably reached: any system of tools and energy will eventually reach its zenith and, thereafter, the system becomes static.

A refinement in the use of energy and greater technological efficiency results in an enhanced ability to conduct the work of habitat transformation. In this way, the developments that White encapsulates in his model may eventually result in a level of resource extraction and in the production of waste, that may impose limits that cannot be breached without, in extreme circumstances at least, risking global catastrophe [4].

Viewed from the perspective of our discussion of limits, the scientific inspection of which started in the eighteenth century and subsequently influenced ecological, evolutionary, demographic, and economic debates, the Anthropocene can be seen as the latest manifestation of limits (but now conceived on a global level). The conjecture that we are bumping up against important planetary limits, the breaching of which can only result in devastation for humans and for the rest of nature, is one of the key notions that frame our understanding of the contemporary world. Models that assess the degree to which boundaries have been breached, or, in some cases, that we are rapidly approaching is the subject of the next chapter.

References

1 Catton, W.R. (1982). *Overshoot: The Ecological Basis of Revolutionary Change.* University of Illinois Press.

2 Schrodinger, E. (2012). *What Is Life?: With Mind and Matter and Autobiographical Sketches.* Cambridge University Press.

3 Daly, H.E. (1996). *Beyond Growth: The Economics of Sustainable Development.* Beacon Press.

4 Meadows, D.H., Meadows, D.L., Randers, J., and Behrens, W.W. (1972). *The Limits to Growth; a Report for the Club of Rome's Project on the Predicament of Mankind.* New York: Universe Books.

5 Hairston, N.G., Smith, F.E., and Slobodkin, L.B. (1960). Community structure, population control, and competition. *The American Naturalist* 94 (879): 421–425.

6 Worster, D. (1994). *Nature's Economy: A History of Ecological Ideas.* Cambridge University Press.

7 Darwin, C. (1996). *The Origin of Species: Oxford World's Classics.* Oxford University Press.

8 Malthus, T.R. (1992). *Malthus: An Essay on the Principle of Population.* Cambridge University Press.

9 Herbert, S. (1971). Darwin, Malthus, and selection. *Journal of the History of Biology* 4 (1): 209–217.

10 Harper, J.L. (1967). A Darwinian approach to plant ecology. *The Journal of Ecology* 55 (2): 247–270.

11 Wallace, A.R. (1871). *Contributions to the Theory of Natural Selection.* Macmillan and Company.

12 Cohen, J.E. (1996). *How Many People Can the Earth Support?* WW Norton & Company.

13 Gotelli, N.J. (2008). *A Primer of Ecology.* Sunderland, MA: Sinauer Associates.

14 Hayes, B. (1993). Computing science: balanced on a pencil point. *American Scientist* 81 (6): 510–516.

15 Godwin, W. (2015). *Enquiry Concerning Political Justice: And its Influence on Morals and Happiness.* UK: Penguin.

16 Petersen, W. (1971). The Malthus-Godwin debate, then and now. *Demography* 8 (1): 13–26.

17 Darnell, R. and White, L.A. (2004). *Biographical Dictionary of Social and Cultural Anthropology* (ed. V. Amit). Routledge.

18 White, L.A. (1943). Energy and the evolution of culture. *American Anthropologist* 45 (3): 335–356.

19 White, L. (2006). Energy and tools. In: *The Environment in Anthropology: A Reader in Ecology, Culture, and Sustainable Living* (ed. N. Haenn and R.R. Wilk), 139–144. New York University Press.

20 Clutton-Brock, J. (1999). *A Natural History of Domesticated Mammal.* Cambridge University Press.

21 White, L. (1975). *The Concept of Cultural Systems: A Key to Understanding Tribes and Nations.* New York and London: Columbia University Press.

22 Barrett, R.A. (1989). The paradoxical anthropology of Leslie White. *American Anthropologist* 91 (4): 986–999.

10

Modeling the Limits

There is a growing acceptance in environment disciplines and in the field of stratigraphic geology that we are now living in the epoch of the Anthropocene, the proposed name for the present stratigraphic moment dominated by human activity. To recognize this is to acknowledge that the anthropogenic disordering of the environment is now manifesting at a *global scale* and is having *global repercussions*. Placing the lower boundary for this new stratigraphic epoch with the transition to the use of fossil fuels as a primary energy source for industrial development acknowledges that our commitment to this energy source represents an important breach with the past. The modification of the atmosphere – one of a number of *environmental sinks* for the waste produced by fueling our economic activities – was decisive, and even before we knew that this was occurring, the chemistry of the atmosphere was subtly changing. This alteration of the composition of the atmosphere increased the risk that our activities would change climatic processes. New climatic conditions alert us to the possibility that humans have, in a decisive way, *exceeding the limits* of the planet's ability to support habitual (business-as-usual) economic and extractive practices.

Though we might have an *intuitive* sense that planetary limits are now breached, how might we quantitatively establish whether this is, or is not, the case? In this chapter, we explore the lines of evidence suggesting that human activity is now bumping up against the limits of the planet, or, indeed, may have already seriously gone beyond the bounds. The pages that follow are quite dense with descriptions of various conceptual models that help in think about limits and boundaries. Once this information is mastered, you will be in a better position to assess the extent to which we are taking unwise risks with the long-term future of both the human enterprise and with the health of the global ecosystem.

At the time of writing, the world population stands at 7.9 billion people. This is certainly a very large number, though all humans, standing side-by-side, could theoretically be squeezed into a single US State. However, as living beings, we need more than mere standing room: we need space to accommodate our resource needs, and space to provide sinks for our waste. Collectively, almost eight billion people cast a shadow over a large amount

A Primer on Human Impacts on the Environment: The Conceptual Approach, First Edition. Liam Heneghan.
© 2023 John Wiley & Sons Ltd. Published 2023 by John Wiley & Sons Ltd.

of ecologically productive land. Is there land sufficient to our needs with enough remaining for non-human creatures? How close are we to the limits of Earth? *Source:* Esaias van de Velde / Wikiart / Public Domain.

10.1 Good Anthropocene/Bad Anthropocene

Although some researchers argue for setting the date for the onset to the Anthropocene quite early in history, the fact that most commentators situate the boundary as recently as the nineteenth century indicates that this is a relatively recent departure from the geological past. The designation of the Anthropocene implies that humans have a worldwide impact but this *does not,* in itself, indicate that we have reached a limit to growth (demographically, economically or culturally). Since the commencement of the Anthropocene there has been a vast expansion in the human use of stored energy, an increase in the exploitation of material resources, and a concomitant elevation in the production of wastes [1]. Quite clearly an acceleration of the human enterprise has been possible even during this new geological epoch. The conspicuous expansion of economic activity and a growing awareness of the environmental impacts has, nonetheless, provoked a disconcerting feeling that something about the world is curiously awry. Regarding the uncanny impression that we has exceeded important thresholds, and that something important has been broken, writer Bill McKibben, (who penned one of the first, generally accessible, books on climate change on – *The End of Nature* [1989]), wrote

> An idea can become extinct, just like an animal or a plant. The idea in this case is "nature"— the wild province, the world apart from man, under whose rules he was born and died. We have not ended rainfall or sunlight. The wind still blows—but not from some other sphere, some inhuman place. It is too early to tell exactly how much harder the wind will blow, how much hotter the sun will shine. That is for the future. But their meaning has already changed [2].

An acceleration of human social-economic activity has undoubtedly occurred over the past half century, and this continues unabated; without a doubt it has placed a strain upon on global ecosystems [1, 3]. But for all of this, it might still be possible to expect, as some interdisciplinary environmental thinkers do, that economies can continue to develop, albeit in a modified form, on this planet radically transformed by human activity. The environmental scientist Erle Ellis, in trying to frame a constructive response to the challenges of the Anthropocene has written:

> Creating that future will mean going beyond fears of transgressing natural limits and nostalgic hopes of returning to some pastoral or pristine era. Most of all, we must not see the Anthropocene as a crisis, but as the beginning of a new geological epoch ripe with human-directed opportunity. [4]

This optimism is the foundation of bids to develop a so-called "good Anthropocene." Exploiting opportunities for social and economic innovation and development – ingenuity in the face of the environmental challenges of our new epoch – can be regarded as "seeds" for the good

Anthropocene [5, 6]. The emergence of the good Anthropocene might still be regarded as "growth" but this is growth understood as innovation, adaptation, and community advancement, rather than merely as continued exploitation of resources. There may undoubtedly be many opportunities in the future, but the rapid (and recently explosive) development of several large regional economies along exploitative lines, in the view of several researchers, has now breached several important planetary thresholds. One way or another, the meaning of the Anthropocene is that the future will not look like the past: we are, other words, facing into a no-analogue future [7]. In the absence of environmental remediation of such problems and without a genuine rethinking of economic models there may be little prospects of a "good" Anthropocene: all options will be ugly [8, 9].

10.2 The Great Acceleration

In the last chapter we discussed the energy and culture model of Leslie White, in which cultural development is understood to be directly related to the amount of energy harnessed by humans *or* by increases in the efficiency that can be realized by a system of technology. This relatively simple model describes at least some of the dynamics of the Anthropocene quite successfully. As we have seen, the initial architects of the idea of the Anthropocene argued for setting the lower bound at the start of the industrial era. Thus, in terms familiar from White's model, the Anthropocene coincides with a new phase in cultural development – that is, the Anthropocene names this new phase.

Though the use of stored carbon as a fuel source dates back into antiquity, the exploitation of fossil fuels and the emerging technologies of the industrial era opened up a vast suite of transformative possibilities. Echoing the Victorian anthropologist Edward Taylor, the historian and philosopher of technology Lewis Mumford wrote as follows in his influential book entitled *Technics and Civilization* (1934):

> The great shift in population and industry that took place in the eighteenth century was due to the introduction of coal as a source of mechanical power, to the use of new means of making that power effective — the steam engine — and the new methods of smelting and working up iron. Out of the coal and iron complex a new civilization developed [10].

This "new civilization" growing out of the steam engine and related innovations resulted in novel possibilities for human communities. It also heralded a massive acceleration in the uses of resources and in human population growth. Recent attempts to quantify changes in the resource relationship between humans and the global environment that started with the industrial era have documented this increase in resource use since the late nineteenth century. However, an even swifter acceleration in the human impacts has occurred since the end of the Second World War (1945 to present). In remarking on this unprecedented growth in the past 75 years, chemist and climate scientist Will Steffen together with his colleagues wrote, "One feature stands out as remarkable. The second half of the twentieth century is unique in the entire history of human existence on Earth" [11].

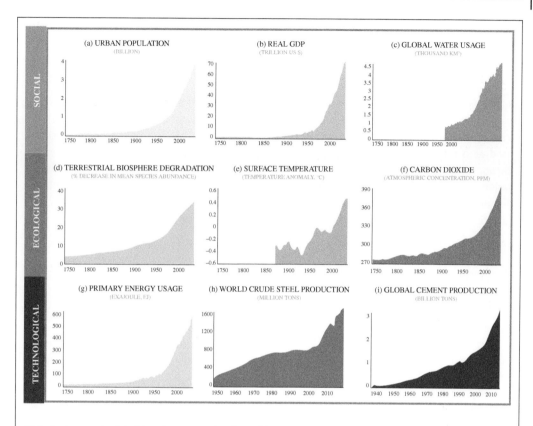

Global urban and related trends are accelerating with cross-cutting impacts on human and earth systems. a: Global urban population data. b: Global real GDP (Gross Domestic Product) in year 2010 US dollars. c: Global water use is sum of irrigation, domestic, manufacturing, and electricity water withdrawals from 1900 to 2010 and livestock water consumption from 1961 to 2010. d: Percentage decrease in terrestrial mean species abundance relative to abundance in undisturbed ecosystems as an approximation for degradation of the terrestrial biosphere. e: Global surface temperature anomaly. f: Carbon dioxide from fern and ice core records. g: World primary energy use. 1850 to present based. h: Crude steel production in 66 countries globally. i: Global cement production. Plot styles are adapted from Steffen et al, 2015 (see original publication for data source [3]. Graph and caption shared under Creative Commons, Open Access).

The "Great Acceleration," (GA), the term used by Will Steffen and colleagues to recognize this amplification of human resource use and impacts, aims to capture the all-encompassing nature of the socioeconomic and biophysical changes that have occurred post-1950. Illustrating the enormity and the pace of the GA are a set of graphs depicting patterns in the social data and changes in earth processes observed over the same duration. Population increases, economic growth, and expansions in resource use, urbanization, globalization, transport, as well as in communication systems are some of the representative social changes; the earth system indicators include altered atmospheric composition, stratospheric ozone, and the climate system, as well as in the water and nitrogen cycles, in marine and land-based ecosystems, loss of tropical forests, and the ongoing degradation of the terrestrial biosphere [3].

A *hole in the ozone layer* – a layer of ozone (trioxygen, O_3) that absorbs harmful ultraviolet radiation from the Sun – was detected in the 1980s. The cause of this thinning (referred to as a hole) was attributed to the manufactured chemical chlorofluorocarbons (CFCs), which are used as a refrigerant and for other purposes, that had been released into the atmosphere. Recognizing the severity of the problem – and a recognition of its possibly catastrophic implications – an international treaty (the Montreal Protocol, 1987) established a timetable for diminishing CFC emissions. Though CFCs were scheduled for complete elimination by 2000, full compliance has not been achieved.

Graphs illustrating the GA show some evidence of the acceleration of some economic and earth systems metrics slowing close to 2010. These changes reflect the financial collapse of 2008–2009, though there is now evidence of a resumption of the accelerating pattern after a global financial recovery. The long-lived greenhouse gases – carbon dioxide, methane, nitrous oxide – which are the most significant drivers of climate change, all increased over this period, though methane concentrations slowed somewhat recently (though they may once again be showing signs of reacceleration). There graphs also show a stabilization of the Antarctic ozone hole, a deterioration that had largely stemmed from the release of ozone depleting substances, many of which are no longer in use. There is evidence of a stagnation in marine fish capture, derived, seemingly, from the reality of depleted fish stocks. Because of the continued intensification of agriculture, there is a concomitant reduction in the need for "new" land (since land under cultivation is being made more productive). In addition, the conversion to domesticated lands (urban areas, croplands, and so on) is slowing down.

The GA graphs that illustrate an overall acceleration in both socioeconomic factors and in earth system metrics which have persisted for decades reveal that these factor are tightly correlated; however caution needs to be exercised in interpreting these associated patterns (correlation, as we are often reminded, should not provoke us to infer causation). In most cases, however, conjectures about the relationships between specific social and biophysical trends are supported by well-studied hypotheses. For example, the connection between energy use by humans and carbon dioxide concentration in the atmosphere has been well characterized: as carbon-based fuel is combusted, there is a concomitant elevation in of carbon in the atmosphere.

When initially presented, the evidence for an acceleration of the social drivers and the environmental impacts conjectured to be associated with them were presented as global averages (that is, they did not take account of regional differences in acceleration). This "lumping" has the effect of masking some important patterns: emissions and impacts from more developed economies are much greater than the modest impacts of many less developed economies [12]. Acceleration of the earth system metrics (carbon, water use, and so on) are driven mainly by developed economies (OECD countries). Identifying inequalities in consumption and production of waste, is important for devising policies that are equitable and just [13]. Recognizing the importance of these disparities, recent presentations of GA more clearly distinguish the global disparities in the drivers of resource use and environmental impacts.

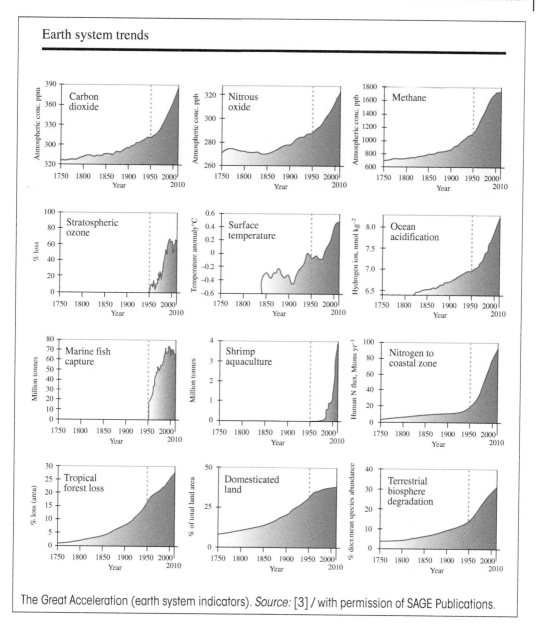

Earth system trends

The Great Acceleration (earth system indicators). *Source:* [3] / with permission of SAGE Publications.

OECD stands for The Organization for Economic Co-operation and Development. It is a multi-national organization founded in 1961 to stimulate economic progress and world trade. The organization helps develop policies to address environmental problems and sustainably manage natural resources. Their reporting explicitly examines linkages between socioeconomic factors and environment outcomes (see https://www.oecd.org/environment).

An important implication of the analysis of GA patterns is that they underscore that even though human alteration of the global environment has been going on for millennia, the scale and rate of change has been greatly amplified in recent decades. Transitioning into the Anthropocene, in itself, may not signify that global limits have been dangerously breaching – but the question, now, is whether this largely undiminished growth spurt is pushing the social-ecological system into a dangerous state with respect to the limits of the earth. How long can the phenomenon of acceleration persist? Will the accelerated exploitation of resources and the modification of planetary sinks ultimately provoke economic decline or collapse, a catastrophic diminishing of some resources (marine stocks, for example), and/or a radical, and detrimental shift in the functioning of key ecosystem processes? Have decades of accelerated change resulted in humanity reaching the limits of the planet?

10.3 Modeling the Limits

To answer the question – "have we reached the limits of the planet?" – we will now turn to a range of models that use data on the social-ecological patterns of the last few decades and project the future states of the world.

The concern expressed in the limits to growth models is that a growing population with immense resource needs will eventually place such a strain on the planet that important limits will be breached. The standard run of the World 3, revealing resources shortages, food production and a declining population. See more details in [14]. *Source:* © 1950 David Kakabadze.

One the most influential environmental models of the past half century examined the implications of continued human population growth and the expansion of the global economy, both considered in the context of planetary limits. It was developed in 1972 by systems scientist and environmental activist Donella Meadows and her team at Massachusetts Institute of Technology (MIT). The computer model used by this team was based upon the pioneering work of mathematical modeler Jay W. Forrester [15]. These are called *World3* models and using them, Meadows' team generated 10 scenarios (later updated to 12) that projected a variety of world futures, based upon a wide range of initiating conditions.

The results of these scenarios were summarized in a widely read and highly influential report: *TheLimits to Growth (LtG)* (1972) [16]. These models were updated three decades later – and published under the unsurprising but ominous sounding title *Beyond the Limits: Confronting Global Collapse, Envisioning a Sustainable Future* (1992). These newly run models largely confirmed and, in some instances, even intensifying the conclusions from earlier model outputs [17, 18]. *The LtG* project and its follow-up can be seen as an important successor to Malthus' model (seen in the last chapter), which predicted an inevitable clash between population growth and the capacity of the earth to furnish resources. The difference in Meadows' work from Malthus's early model is that the input for World3 models goes beyond simply conjecturing a relationship between the growth of the means of subsistence and the growth of the human populations, and, instead, include a wide range of social and ecological variables.

Donella Meadows (March 13, 1941-February 20, 2001) and the team that developed and contributed to the Limits to Growth (1972). The results of the models that Meadows and her team ran illustrated a range of potential future scenarios for our global economy and environment. The most pessimistic of these scenarios foresaw an exhaustion of nonrenewable resources and a decline in the quality of human; the more optimistic models, where drastic changes to economics are instituted results in a sustainable standard of living. The Academy for Systems Change.

World3 models incorporate interactions between the food system, the industrial sector, human population change, the availability of nonrenewable resources systems, and the output of pollution from the economy. The "standard run" of the model, against which all other World3 scenarios are compared, makes predictions about a likely future where no deviation has occurred from business as usual (i.e. it predicted an uninterrupted extension of global acceleration, to use contemporary language). The model predicted a collapse of both the human population and the output of the industrial sector after a period of accelerated growth. This collapse is associated, ultimately, with a depletion of natural resources. As industry grows, more capital is required to seek and extract additional resources, though, eventually, the cost of re-investment in resource exploitation becomes prohibitive. A diminution of food supply drives the death rate and, as a consequence, the population crashes. Even if the resource supply is assumed to become significantly larger – that is, the overall size of the economy increases – nonetheless, all else being equal, collapse is inevitable (at most, it is forestalled for a few years). Diminishing food supply and high pollution levels will always lead to a sharp decline in human population [17]. The authors concluded the "the limits to growth on this planet will be reached sometime within the next one hundred years" [16]. The model (and the book based upon it) generated considerable media and academic attention.

Though the authors of *The LtG* report were more than willing to acknowledge some of the shortcomings of their model, nevertheless they professed considerable confidence in the robustness of their conclusions and predictions, and redoubled that confidence once they had replicated their work 30 years after they had performed the initial model run [16]. Despite the confidence of the authors, there has, however, been considerable criticism of the World3 approach [19]. The critiques of World3 center on a number of key points.

1) Some critics have been skeptical about *the quality of the assumptions* regarding the mechanisms connecting parameters in the model. Large knowledge gaps existed at the time about how subsystems are connected and this deficit may be sufficiently severe so as to undermine the conclusions of the project. According to Janssen and De Vries, the relationship between, for example, pollution and health, as well as the putative interactions between demographic factors and economic dynamics were poorly understood at the time the model was developed. Furthermore, the significance of technological innovations for increasing resource availability was not fully apparent in the 1970s (and, of course, the prospect for techno-solutions to resource problems remains contestable to this day) [19].

2) There was a *paucity of reliable data* available for model calibration and validation when the first of the world-models was presented. The model was thus seeded with data that were, at best, fragmentary.

3) The computer *model itself employs a high aggregation level*, i.e. it considers the world as a whole, rather than examining regional patterns. (The problem of aggregation is one that we already discussed in relation to the more recent Global Acceleration data). A more disaggregated model might reveal important and complex patterns at higher degrees of geographical resolution [20, 21].

Model aggregation refers to the way in which the data in the model are compiled from several smaller pieces that form a larger whole. This observation, you will recognize is rooted in basic systems theory). Models can combine blocks of information is a variety of ways: for instance, in social-ecological models, the model can aggregate by country, voting blocks, economic regions, and so on. Lower resolution analysis must be used if researchers want to pinpoint the source of the primary drivers of larger scale changes in a global system.

4) Because of poorly delineated hypotheses connecting the subsystems of the model, World3 is accused of defaulting to *unduly pessimistic* scenarios. The pessimism derives, in large part, or so its critics claim, from the subjective estimates of the world's carrying capacity upon which the models rely. In their harsh critique of the World3 model, Pavitt and colleagues, in a book entitled *Models of Doom: A Critique of The Limits to Growth* (1973) [22], suggested that the Meadows' project was based upon faulty Malthusian conclusions. As Christopher Freeman wrote in that volume "'Malthus in, Malthus out.' . . . what is on the computer print-out depends on the assumptions which are made about real-world relationships, and these assumptions in turn are heavily influenced by those contemporary social theories and values to which the computer modelers are exposed" ([19, 22, 23]).

Many of these initial criticisms of the World3 model are cogent and undeniably hit their mark. As a result of continuing criticism of these models, there is widespread caution about the conclusions of *The LtG* project, especially among technological optimists [24]. However, the essence of the critiques against *The LtG* are founded on claims about the paucity of both data and mechanistic hypotheses that informed the model and, since much more data have become available, these objections now seem considerably less compelling. It is also clear from the subsequent writings of the authors of *The LtG* that the World3 model were never intended to provide detailed forecasts about the future of the world. Rather these models provided a *conceptual framework* for thinking about possible outcomes of a particular form of economic growth. These were among the first of such global environmental models and were designed primarily to provoke more profound efforts to gain a clear understand of the behavior of the world economic system in relation to the resources that sustains it. Thus, the various scenarios run in the World3 models were not explicitly predictive (or, at least, this is what the authors often claim); rather, they were attempts to envision economic development and population growth under a variety of conditions.

As we have noted, in the decades since the initial publication of *The LtG*, considerably more data have become available on demographic trends, environmental impacts, on the growth in the agricultural and industrial sector, and on the availability of nonrenewable resources. When the performance of the "standard run" from 1972 was re-assessed, but this time by a research team not associated with Meadow's original tem, data available from 1970 to 2000 was used. Once again the standard model continued to perform reasonably well, and, this updated model also forecasted global trouble in the near future [24]. The future that the model predicted was largely confirmed, even using considerably better data. Troublingly, business as usual scenarios continue to anticipate a global collapse before the middle of the twenty-first century [24, 25].

10.4 Calculating the Global Ecological Footprint

The conclusions from *The LtG* modeling project are grim. If the conditions pertaining to the standard run continue unobstructed, then economic and demographic decline in the coming decades seem, at the very least, plausible. For all of that, the criticisms of the World3 models' assumptions and the recognition that the projected scenarios were initially based on limited data should alert us to the need for a variety of tools to allow us to think clearly about the future. What other tools are available for estimating the limits to growth?

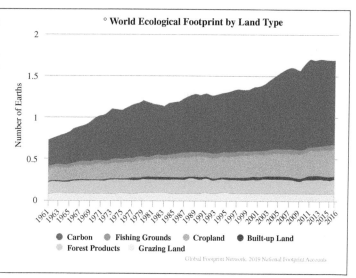

Ecological Footprint (EF) from 1961 to 2016. EF is broken down by the demand for different functional categories of the land. The units of the dependent axis illustrate the number of earths required to sustain the collective human population.

Another approach to addressing the question of global limits is represented by the development of resource accounting tools, of which "ecological footprinting" (EF) is the best known. EF determines whether the collective demand for resources by the world's economies remains within the limits of the planet's natural capital stocks. The earth's "biocapacity" is the preferred terminology used by the Global Footprint Network (GFN) to refer to our planet's collective capital stocks [26–28]. The concept of the EF was developed by ecologist and public policy professor William E. Rees and his then doctoral student Mathis Wackernagel, a specialist in regional planning and sustainability studies. At the core of this concept is the recognition that in order to sustain human life and well-being a considerable amount of land is required. The footprint of human habitation, our cities, and so forth, is merely the most conspicuous land occupied by humans. To assess our full ecological footprint requires a complete reckoning of "all the biologically productive lands required to sustain life and adsorb waste using prevailing technology and resource management practices" [29]. The biocapacity of ecosystems that is needed to both supply and to regenerate services is expressed in a normalized measure of land area called global hectares (abbreviated as "gha"). The curious standardized measurement, the global hectare, takes account of the fact that different land types possess different levels of productivity. A global hectare thus represents the world's average biological productivity for a given year. To clarify this idea: imagine the difference in the productivity of rich farmland compared to considerably less productive pastureland: a global hectare of the former will be smaller than the later. In this way the biocapacity expressed in global hectares allows for standardized comparisons between different countries and regions [30].

An ecological footprint can be calculated on scales that range from individuals and countries, and all the way up to the global scale. Since the calculations rely exclusively on publicly available data, they can be replicated and assessed independently of the investigators who invented the analysis [26]. Therefore, if you are skeptical about the conclusions of EF analysis (which like the LtG models are also somewhat grim), you can and should, play around with the data.

At a regional scale the dimensions of the required ecological footprint depend on population size, the technological systems being employed to maintain the economy, the prevailing material standards of living, and the ecological productivity of the lands from which resources are drawn and wastes are sequestered [27]. Scaled up to the level of the globe, this resource accounting technique permits an evaluation of the *capacity of the entire globe* to support (or, perhaps not) the global human population without exceeding the *regenerative capacity* of global ecosystems. The reference to the regenerative capacity of ecosystems is an important one. The term underscores the fact that it may be possible, at least in the short term, for the human population to exceed the capacity of the area of land required to sustain it (and this circumstance is very likely to be the case, globally). By drawing down of the capacity of land (its natural capital), a population eventually exhausts the regenerative ability of ecosystems. Humans can overshoot the limits of the planet – that is, place demands on nature that exceed supply – only because it is possible to draw down "natural capital," the ecological assets that continue to provide goods and services [31, 32]. In such circumstances continued growth is sustained by exhausting the capacity of the land to replenish supply. Such as system is doomed to eventual collapse.

The concept of **overshoot** was systematized by the sociologist William Catton in his book *Overshoot: The Revolutionary Basis of Ecological Change* (1980). Catton defines it as the capacity of a population to ". . . increase in number so much that the habitat's carrying capacity is exceeded by the ecological load, which must in time decrease accordingly; (n.) the condition of having exceeded for the time being the permanent carrying capacity of the habitat."

According to Wackernagel and his colleagues writing in a paper published in 2002, "Tracking the ecological overshoot of the human economy," the demand for biocapacity has exceeded the entire biosphere's regenerative capacity since the 1980s [32]. In that paper, they observe:

> According to this preliminary and exploratory assessment, humanity's load corresponded to 70% of the capacity of the global biosphere in 1961, and grew to 120% in 1999. By 2014 it was estimated the footprint was at 170% (or 1.7 Earths) and continues to grow [33].

More recently, EF has been used to assess *the annual onset* of "overshoot." Every year the Global Footprint Network (GFN) – a think tank of which Wackernagel is founder and president – calculates an earth overshoot day (EOD) – the calendar date when humans have breached the annual threshold of biocapacity for that year [34]. In 2020, for example, EOD fell on the 22 August: from that date on to the end of the year, humans draw down the capital stocks of the planet (https://www.footprintnetwork.org). In 2021, the EOD fell on July 29. To illustrate how dramatically humanity is digging into its reserves, it is worth noting that in 1970 human needs were met, more or less, within the regenerative capacity of the earth (that year EOD fell on December 30; we lived for one day on borrowed capacity).

10.5 Limitations of Ecological Footprinting

EF is used as a metric to assess sustainability progress in several municipal and even national arenas [35–37]. Although EF, as a tool for accounting of resource use has stimulated widespread exploration of environmental limits, it has also garnered considerable critique from the scientific community [38–41]. For example, in their sharply worded rebuke of the concept of EF entitled "Footprints to Nowhere" (2014), Mario Giampietro and Andrea Saltelli warned:

> Crisp numbers make it to the headlines. However, it is unlikely that a single crisp number can capture a complex issue, such as the analysis of the sustainability of human progress both at the local and the global scale. [40]

The concept of **sustainability** became prominent in environmental thought after the release of the Brundtland Report "Our Common Future," in 1987. In this report, *sustainable development* was defined as "development that meets the needs of the present without compromising the ability of future generations to meet their own needs" (Report of the World Commission on Environment and Development: Our Common Future. United Nations General Assembly document A/42/427.) [42].

Giampietro and Saltelli accuse EF of being "media friendly" and argue that "EF's rhetoric trivializes bio-economics and muddles the sustainability debate" and, furthermore, that from a policy perspective "the existing EF is useless and misleading" [43]. Other critics had previously articulated their concerns over the policy aspects of EF [44]. Of particular concern is that EF reduces resource use (sources) *and* waste production (sinks) to a single metric – that is, the footprint is expressed in global hectares – and this single measurement is used to assess the highly complex issue of sustainability [40, 45]. Sustainability cannot be regarded merely as a problem of satisfying human demand, therefore, ideally, relevant measurements should include an assessment of the state of biodiversity, human health and well-being, community welfare, social equity and an evaluation of nature protection, in addition to a range of other properties [46].

Substantial critiques of the science behind EF center on queries about the computational aspects of its resource accounting methods: the protocols are, its critics claim, cumbersome and the outputs should be treated circumspectly (they are "ultimate fragility") [40]. Both Giampietro and Saltelli, and another team led by Linus Blomqvist from the *Breakthrough Institute* who, along with his colleagues, express a concern that because of the way in which biocapacity is evaluated (and converted into global hectares), most of the "overshoot" measured in EF calculations comes from carbon footprint [38]. According to Blomqvist, there has been little change in the required footprint of cropland, forests, grazing land, marine and inland fishing grounds, and infrastructure. Blomqvist and colleagues express their concern in the following way:

> When the global EF is decomposed into its six components, none of the five non-carbon land-use categories has any substantial ecological deficit – suggesting that depletion of cropland, grazing land, forest land, fishing grounds, and built-up land is not occurring on an aggregate, global level. This result stems from the fact that the accounts for cropland, grazing land, and built-up land are constructed in such a way that they are always near equilibrium, with the footprint of consumption by definition nearly equal to biocapacity; fishing grounds and forest land are both in surplus. Hence, virtually all of the ecological overshoot comes from the EF's measure of the rate at which carbon dioxide is accumulating in the atmosphere. [38]

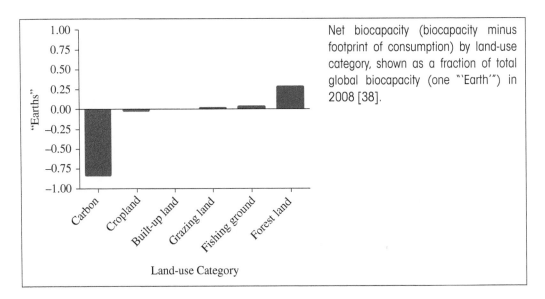

Net biocapacity (biocapacity minus footprint of consumption) by land-use category, shown as a fraction of total global biocapacity (one "'Earth'") in 2008 [38].

To understand the essence of this criticism, consider cropland: both production of and demand for food ideally should become almost equal. Humans eat what they grow: though significant quantities of what is grown are lost to spoilage and to pests (an estimated 7% of annual cropland acreage is lost to waste [47]). However, over time, both demand (as population grows) and production (ability to satisfy calorific needs) can increase without this elevation of production and consumption necessitating an increase in the area of land under production. This is because agriculture can become more intensive through increased use of pesticides (to reduce pests), fertilizers, enhancements in crop and animal breeding, reduced wastage, and so on. Since the production of both fertilizer and nitrogen are energetically expensive, the capacity to increase production in croplands is, therefore, reflected not in the increase in cropland footprint, but by means of an increase in carbon footprint (that is, in the need for biologically productive area necessary for absorbing this CO_2). Furthermore, it is worth considering that the environmental impact of agriculture and its associated energy needs (and carbon emissions associated with it) include soil erosion, phosphorous and nitrogen leaching into water, and a loss of biodiversity [39]. Such increases in impact are, at best, only ambiguously captured by the ecological footprint.

In response to this specific criticism, the GFN acknowledges that biocapacity can be increased over time with, for example, the use of pesticides and fertilizers, as we have already noted, and through the implementation of a variety of management practices [41]. Such practices can be detrimental in the long run and these detriments are not fully captured in the calculation of a footprint for cropland. Rather, they may be reflected in a country's carbon footprint (this fact recognizes that fertilizers and pesticides are energetically expensive to produce and that energy often comes from the burning of fossil fuels). However, when comparing among different managements (intensive versus extensive farming, for example) the EF method, according to its advocates, "does not judge which option is preferable; it simply sheds light on the resource flows associated with each one" [41].

Furthermore, GFN acknowledged that its computational approach has evolved over time, and that in many cases its critics have assailed older versions of its methodology [33, 48]. Furthermore, it argues that the approach it takes to resource accounting – balancing human demands against the capacity of nature to meet that demand –, does not in itself prescribe any particular policy solutions. Rather it provides data that may be useful in developing sustainability strategies – it may not be the only information needed for instituting sustainability strategies, but it is helpful information nonetheless [41].

Critics of EF suggest a range of alternative measures for sustainability that might prove more meaningful for the development of policy; these include, as examples:

1) *The Planetary Boundaries* (PB) model, which seeks to estimate "the safe space for human development" [49].
2) *The Nature Index*, a framework that draws from a number of sources to assess biodiversity, and presents this information in an accessible form for managers, policy makers, and the public [50].
3) *Ecosystem Integrity, Multi-Scale Integrated Analysis of Societal and Ecosystem Metabolism* (MuSIASEM), which evaluates patterns in the metabolism of socioeconomic systems on a range of scales from households to the global economy and characterizes these in terms of the performance of socioeconomic activities and ecological constraints by comparing matter and energy flows under the influence of human activity compared to the patterns found in natural ecosystems [51].

Each of these methods can usefully provide information on different aspects of sustainability for practitioners and policy makers. However, of these, the first, that is, the PB model, is highly relevant to the question being asked in this chapter – are we living safely within the limits of the planet? So we will examine this model in some detail.

10.6 Living in a Safe Place: Planetary Boundaries

The **precautionary principle** urges an exercise of restraint as a principle to protect the environment in the face of an *incomplete understanding* of the impacts of actions and products. The principle derives from the United Nations Charter for Nature published in 1982 and it is incorporated into a variety of local, national, and international policy agreements. Very often policy makers must intervene to reduce risk even if the scientific evidence on a topic is incomplete.

The PB approach extends the framework developed by Donella Meadows and reported in the *LtG* reports and books. Like ecological footprinting, PB investigates limits by taking seriously the concerns that if business-as-usual models persist then humans face a potentially catastrophic future [49, 52]. However, in addition to extending the thinking of *The LtG*, PB also draws upon a variety of other important antecedent concepts. These include the precautionary principal and notions of "tolerable windows" (this approach is based on a process of identifying a set of "guard-rails" beyond which climate change, for example, cannot go) [53, 54]. The PB approach addresses the question of whether the human enterprise can be safely contained within the capacity of planetary processes to sustain it. The approach also incorporates the insight that breaching a threshold risks producing nonlinear biophysical behaviors in the system – outcomes that may be substantially larger in scope than expected – and that these outcomes may drastically undermine the conditions required for economic prosperity [49]. Johan Rockström and colleagues, who pioneered this model, phrase their concern about the implications of these *critical transitions* as follows:

> Thresholds are intrinsic features of those systems [coupled human–environmental systems] and are often defined by a position along one or more control variables, such as temperature and the ice-albedo feedback in the case of sea ice. Some Earth System processes, such as land-use change, are not associated with known thresholds at the continental to global scale, but may, through continuous decline of key ecological functions (such as carbon sequestration), cause functional collapses, generating feedbacks that trigger or increase the likelihood of a global threshold in other processes (such as climate change) [49].

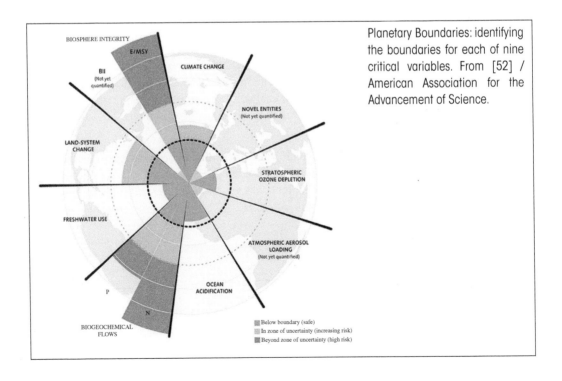

Planetary Boundaries: identifying the boundaries for each of nine critical variables. From [52] / American Association for the Advancement of Science.

What is meant by this statement is the following. Some important environmental variables, such as those nutrients in soil or aquatic systems that affect the fertility of that system (to add to the examples that Rockström and colleagues provide in the quote above), may fluctuate in availability up to a certain point without changes in these variables being reflected in extravagant alterations to the behavior of systems. From the perspective of human welfare, therefore, there are fluctuations that can and do occur – even under the influence of human activity – that will not disrupt the sustainability of the system. However, there also exist *thresholds* beyond which even a modest change in that variable alters the behavior of a system dramatically. The theoretical framework for understanding systems behavior (and misbehavior) is something that we take up later (especially in the final chapter of this book), but here we should note, based upon the example of nutrient limitation in soil or water that the *notion of a limit* in the PB literature that is quite different from limits as understood in its conceptual predecessors. Whereas in the *LtG* model and EF analysis, limits refer to the capacity of the planet to provide resource quantities needed to sustain the growth of human enterprise and by providing adequate human sustenance or by providing sufficiently capacious sinks for waste, the PB model extends the discussion of limits to include the sort of boundaries that cannot be breached without altering the habitual behavior of systems. Thus, limits in the context of the PB model reflects the necessity that economic development must be constrained within the bounds of the *behavioral stability* of the Earth system.

The term "boundary," as used by the Rockström team mean to prescribe a set of "safe" distances from a dangerous threshold. There may be some uncertainty about precise location of the threshold for a control variable, so a buffer may exist between the setting of a boundary and the behavioral threshold of the system. This buffer reflects an application of the precautionary principle. Setting cautious limits on development (through the application of a precautionary principle) and implementing a warning system based upon the monitoring of important earth system metrics helps avoid risky management decisions [52].

Unlike the EF, where there is a single metric used to guide policy – that is, the footprint – the PB approach is more disaggregated. Potential threats to the benign functioning of the earth system as judged from the human perspective are broken down into nine (intersecting) boundaries, each of which can be monitored and assessed for warning signs.

1) *Climate change* has, in the assessment of many climate scientists, not only transgressed a planetary boundary but is approaching a significant Earth system threshold, provoking dramatic changes in system behavior [55, 56]. The current levels of CO_2 in the atmosphere (the primary control variable considered in addressing climate change boundaries) is 418 ppm compared to "pre-industrial" levels of 280 ppm. An atmospheric level of 350 ppm CO_2 is considered in the safe range, and is often given as the goal for climate mitigation strategies [55]. An evaluation of the record of past conditions suggests that about 550 ppm is associated with an ice-free earth. Physical feedback mechanisms triggered by an ice-free world would exacerbate a rise in temperatures. A still warmer state will have implications for drastic sea level rises. Other thresholds associated with accelerated temperature increases include a loss of carbon sinks (by, for example, a transformation of tropical forest) and would, in turn, intensify climate impacts [57]. The implications of climate change are taken up in detail in a later chapter.

2) *Novel entities*, by which is meant new chemical substances, or novel forms of existing substances, and life forms genetically modified in a variety of ways, may have a range of effects on both abiotic and biotic processes. Over 100 000 of these novel entities have been recognized [52]. The behavior of such novel substances and modified entities at the level of planetary processes is insufficiently understood at the present time, despite the considerable knowledge generated

at more local scales (for example, the ecotoxicological influences on soils or individual organisms, and so on). Therefore, it is not possible to present a generalized set of boundary conditions for safe operation with respect to novel entities. Given these circumstances, the precautionary principle can be invoked to prevent planetary level harm even if the threats are poorly defined. The harm to the ozone layer associated with the release of chlorofluorocarbons (CFCs) – a class of compounds that provided a range of benefits, and that were initially regarded as environmentally benign, but which had, as a consequence, major planetary implications – provide a cautionary tale regarding the large scale thresholds associated with novel entities [58].

3) *Stratospheric ozone depletion*, mentioned above, is represents a planetary boundary in its own right and the earth passed a critical threshold for this substance signified by the appearance of the Antarctic ozone hole in the 1980s [59]. The problem was somewhat unexpected: the release of ozone-depleting chemical substances (CFCs were noted as of particular concern, though they are not the only contributors to reduced stratospheric ozone) resulted in an overall depletion of the planet's ozone layer, but also, at the regional scale, a more dramatic reduction of springtime ozone levels at the South Pole where ozone declined by more than 40% [59, 60]. The altered ozone layer and the resulting increase in exposure of this polar region to ultraviolet radiation had a range of consequences for human health, and for biological and earth system processes (including impacts on productivity of key planktonic species important to marine food webs, and as a contributor to climate change) [61]. As a consequence of these concerns, there was a decisive global policy action plan put in place. The Montreal Protocol, which went a long way toward the elimination of ozone-depleting chemical substances, has brought these substances back to safe levels [62].

4) *Atmospheric aerosols* of environmental significance include particles (with a wide range of diameters) found in high concentrations. Aerosols can influence air quality and a variety of earth system processes. Altered aerosol concentrations can have implication for human health as well as upon the regional climate. They affect climate by altering the absorption of solar radiation, modifying the behavior of clouds and thus reduce the solar irradiance that can reach the Earth's surface. Furthermore, the environmental impacts of aerosols can lead to a modification of the hydrological cycle [63]. Although there is considerable information available about anthropogenic aerosols effects, the implications of elevated aerosols are not understood well enough at a global scale to allow for PB thresholds to be set [49, 52].

5) *Ocean acidification* is likely to have a threshold that defines the limits to the safe operation of some key marine ecosystem processes. However, the boundary position is uncertain as much more needs to be known about the global-scale responses to ongoing acidification [52]. A lowering of ocean pH is associated with rising atmospheric carbon dioxide (CO_2). This has been referred to as "the other CO_2 problem" [64]. Increased acidity of seawater reduces the amount of the available carbonate "building blocks" used by many shell-forming marine organisms from plankton to corals. Alteration to marine food webs as a consequence of acidification may potentially lead to drastic reductions in marine resources in a way that affects the lives of millions of people [57, 64].

6) *Biogeochemical flows* (especially of nitrogen and phosphorous) have breached safe levels, and the mismanagement of biogeochemical processes has presented a range of interrelated and complex problems for humanity. Much of the new reactive nitrogen produced by human manipulation of the nitrogen cycle is emitted to the atmosphere in various forms rather than being taken up by crops. When nitrogen (and other nutrients) are rained out of the atmosphere, the excess nitrogen pollutes waterways and coastal zones. It also accumulates in the terrestrial biosphere. Similarly, a relatively small proportion of the phosphorus in fertilizers applied to food production systems is taken up by plants; much of the phosphorus mobilized by humans

also ends up in aquatic systems. Aquatic systems can as a consequence become oxygen-starved as bacteria consume the blooms of algae that grow in response to the high nutrient supply. A significant fraction of the applied nitrogen and phosphorus makes its way to the sea, and pushes marine and aquatic systems across ecological thresholds of their own. One regional-scale example of this effect is the decline in the shrimp catch in the Gulf of Mexico's "dead zone" caused by fertilizer transported in rivers from the US Midwest.

7) *Freshwater use* boundaries are evaluated in relation to the consumption of blue water (that is, consumption of ground- and surface water flow). 4000 km^3/year global is the value given as the boundary beyond which dramatic changes in the functioning of flow dependent ecosystems occur. More recent assessments of freshwater use also evaluate basin-scale boundaries for water withdrawal along major river courses. Safe levels of use are determined to some extent by measuring the mean monthly flow of rivers [52]. Though the PB team regards the current level of freshwater use to be within safe limits (while also expressing certain equivocation about where this boundary is set)– a 2015 assessment calculated that total fresh water use was around 2600 km^3/year – there has been considerable skepticism by other researchers who argue that recent advances in understanding the dynamics of freshwater use indicates that it is not, in fact, at a safe level (and that use may already exceed the 4000 km^3/year boundary) [65]. The "message of apparent calm" from the PB, these critics argue, should be disregarded.

8) *Land system change* is evaluated for the purposes of planetary boundary analysis to the extent that these changes influence climatically relevant biogeophysical processes, such as the exchange of energy, water, and others that determine the interactions between land and the atmosphere [52]. In contemporary work, the driving variable is considered the area of forested land expressed as percentage of original forest cover [56, 57]. The boundary for land system changes is not, of course, merely dependent upon the quantity, but also on the quality of habitat and its spatial distribution. Analyzed in this way, changes in land use is considered to be in a zone of increasing risk.

9) *Biospheric integrity* is measured in two categories: genetic and functional diversity. The former reflects the diversity of genetically unique material on Earth; the latter metric evaluates the role that living things play in significant earth system processes (that is, it is an expression of the functional traits of organisms in a variety of ecosystems). In the absence of comprehensive information, extinction rate is assessed to evaluate genetic diversity, and biodiversity intactness is assessed using the Biodiversity Intactness Index (BII). Contemporary biodiversity loss is occurring at such unprecedented rates that a threshold for the integrity of the biosphere has been severely breached, a topic taken up in detail later in this book.

10.7 Coda: The Complex Meaning of Limits

The English word "limit," is borrowed from the Old French, *limite* and Latin, *limitem*, to indicate a boundary or the setting of boundaries. It is used in ordinary speech as well as having specialized meaning in mathematics, logic, metaphysics, and geography. In its primary use, it refers to a bound that should not be passed, or beyond which something ceases to be possible or allowable.

As we have seen in this and the previous chapter, environmental disciplines have employed the notion of limits and boundaries in a way that deploys several nuances of its meaning. In the Malthusian model, sustenance (food) sets a limit on the growth of populations, and thus sets boundaries on the potential perfectibility of the human condition. In the frameworks of contemporary ecology, taking their cue from the Darwinian struggle for existence, resource limits are but one of a number of factors that influence the growth rates of populations, and that determine the outcome of competitive interactions in multispecies communities.

Environmental models – from *The LtG* project, initiated in the 1970s, to the EF and PB models of recent decades – conjecture that the certain types of economic development must operate within limits. This is because economic development depends upon the availability of resources and utilizes environmental sinks to absorb wastes. There are boundaries, which cannot long be breached without consequences.

To live in the Anthropocene is to live in an era when humankind has emerged as a globally dominant force. Human action has arguably now ascended to a level like that of like oceanic circulation, the internal heat of the Earth, tectonic forces, or climatic processes, where our decisions can modify the conditions upon which all life depends. This, of course, does not have to be a bad thing – a Good Anthropocene is still possible; however, to breach important thresholds is to risk an unpredictable future. The models that we have examined in this chapter all converge on the idea that the planetary future (and our human future upon this earth) will not look like the past. However, more than this, most models agree that a failure to monitor the vital signs of the earth is to risk a *catastrophic future*. In thinking about these issues, it is important to remind ourselves that the word "limit" is both a noun and a verb. That is, we can *chose to exercise restraint* (that is, self-limit) based upon a risk assessment of key environmental variables. Failure to pay attention to vital signs runs the risk of falling afoul of limits of a sterner sort: the exhaustion of possibility for any livable future.

The next section of the book assesses the credibility of claims that we are about to confront the more severe limits of the planet. Beyond these only catastrophe exists.

References

1 Steffen, W., Crutzen, P.J., and McNeill, J.R. (2007). The Anthropocene: are humans now overwhelming the great forces of nature. AMBIO. *A Journal of the Human Environment* 36 (8): 614–621.

2 McKibben, B. (1989). *The End of Nature*. Random House Trade Paperbacks.

3 Steffen, W., Broadgate, W., Deutsch, L. et al. (2015). The trajectory of the Anthropocene: the great acceleration. *The Anthropocene Review* 2 (1): 81–98.

4 Ellis, E. (2011). The planet of no return: human resilience on an artificial earth. *Breakthrough Journal*. 2: 39–44.

5 Bennett, E.M., Solan, M., Biggs, R. et al. (2016). Bright spots: seeds of a good Anthropocene. *Frontiers in Ecology and the Environment* 14 (8): 441–448.

6 McPhearson, T.M., Raymond, C., Gulsrud, N. et al. (2021). Radical changes are needed for transformations to a good Anthropocene. *NPJ Urban Sustainability* 1 (1): https://doi.org/10.1038/s42949-021-00017-x.

7 Hobbs, R.J., Arico, S., Aronson, J. et al. (2006). Novel ecosystems: theoretical and management aspects of the new ecological world order. *Global Ecology and Biogeography* 15 (1): 1–7.

8 Dalby, S. (2016). Framing the Anthropocene: the good, the bad and the ugly. *The Anthropocene Review* 3 (1): 33–51.

9 Fleming, D. and Chamberlin, S. (2016). *Surviving the Future: Culture, Carnival and Capital in the Aftermath of the Market Economy*. Chelsea Green Publishing.

10 Mumford, L. (2010). *Technics and Civilization*. University of Chicago Press.

11 Steffen, W., Sanderson, R.A., Tyson, P.D. et al. (2006). *Global Change and the Earth System: A Planet under Pressure*. Springer Science & Business Media.

12 Malm, A. and Hornborg, A. (2014). The geology of mankind? A critique of the Anthropocene narrative. *The Anthropocene Review* 1 (1): 62–69.

13 Sze, J. (2018). *Sustainability: Approaches to Environmental Justice and Social Power*. NYU Press.

14 van Dijkum, C., de Tombe, D., and Van Kuijk, E. (1999). *Validation of simulation models*. Amsterdam: SISWO (Netherlands' Universities Institute for Coordination of Research in Social Sciences).

15 Forrester, J.W. (1971). *World Dynamics*. Wright-Allen Press.

16 Meadows, D.H., Meadows, D.L., Randers, J., and Behrens, W.W. (1972). *The Limits to Growth: A Report for the Club of Rome's Project on the Predicament of Mankind*. New York: Universe Books.

17 Meadows, D., Randers, W.W., J., and Meadows, D. (2004). *Limits to Growth: The 30-Year Update*. Chelsea Green Publishing.

18 Meadows, D.H., Meadows, D.L., and Randers, J. (1992). *Beyond the Limits: Confronting Global Collapse, Envisioning a Sustainable Future*. Post Mills, VT: Chelsea Green Publishing Co.

19 Janssen, M. and De Vries, H. (1999). *Global Modelling: Managing Uncertainty, Complexity and Incomplete Information*, 45–69. Amsterdam: Validation of Simulation Models SISWO.

20 Thissen, W. (1978). Investigations into the World3 model: lessons for understanding complicated models. *IEEE Transactions on Systems, Man, and Cybernetics* 8 (3): 183–193.

21 Thissen, W. (1978). Population in the Club of Rome's World3 model. 8 (3): 149–159.

22 Pavitt, K., Jahoda, M., Freeman, C., and Cole, H. (1973). *Models of Doom: A Critique of the Limits to Growth*. New York: Universe Books.

23 Freeman, C. (1973). 1. Malthus with a computer. *Futures* 5 (1): 5–13.

24 Turner, G. (2008). A comparison of the limits to growth with 30 years of reality. *Global Environmental Change* 18 (3): 397–411.

25 Pasqualino, R., Jones, A., Monasterolo, I., and Phillips, A. (2015). Understanding global systems today — a calibration of the World3-03 model between 1995 and 2012. *Sustainability* 7 (8): 9864–1989.

26 Wackernagel, M., Onisto, L., Bello, P. et al. (1999). National natural capital accounting with the ecological footprint concept. *Ecological Economics* 29 (3): 375–390.

27 Wackernagel, M. and Rees, W. (1998). *Our Ecological Footprint: Reducing Human Impact on the Earth*. New society Publishers.

28 Rees, W.E. (1992). Ecological footprints and appropriated carrying capacity: what urban economics leaves out. *Environment and Urbanization* 4 (2): 121–130.

29 Global Footprint Network. Glossary. https://www.footprintnetwork.org/resources/glossary (accessed 16 November 2022).

30 Kitzes J., Galli, A., Wackernagel, M. et al. (2007). A 'constant global hectare' method for representing Ecological Footprint time trends. *International Ecological Footprint Conference* (December 2007). https://www.researchgate.net/publication/242479376_A_Constant_Global_Hectare_Method_for_Representing_Ecological_Footprint_Time_Trends (accessed 16 November 2022).

31 Costanza, R., d'Arge, R., De Groot, R. et al. (1997). The value of the world's ecosystem services and natural capital. *Nature* 387 (6630): 253–260.

32 Wackernagel, M., Schulz, N.B., Deumling, D. et al. (2002). Tracking the ecological overshoot of the human economy. *Proceedings of the National Academy of Sciences* 99 (14): 9266–9271.

33 Lin, D., Hanscom, L., Murthy, A. et al. (2018). Ecological footprint accounting for countries: updates and results of the National Footprint Accounts, 2012–2018. *Resources* 7 (3): 58.

34 Global Footprint Network (2017). Earth overshoot day. https://www.overshootday.org/newsroom/highlights-2017/ (accessed 16 November 2022).

35 Kitzes, J., Galli, A., Bagliani, M. et al. (2009). A research agenda for improving national ecological footprint accounts. *Ecological Economics* 68 (7): 1991–2007.

36 York, R., Rosa, E.A., and Dietz, T. (2004). The ecological footprint intensity of national economies. *Journal of Industrial Ecology* 8 (4): 139–154.

37 Scotti, M., Bondavalli, C., and Bodini, A. (2009). Ecological footprint as a tool for local sustainability: the municipality of Piacenza (Italy) as a case study. *Environmental Impact Assessment Review* 29 (1): 39–50.

38 Blomqvist, L., Brook, B.W., Ellis, E.C. et al. (2013). Does the shoe fit? Real versus imagined ecological footprints. *PLoS Biology* 11 (11): e1001700.

39 Galli, A., Giampietro, M., Goldfinger, S. et al. (2016). Questioning the ecological footprint. *Ecological Indicators* 69: 224–232.

40 Giampietro, M. and Saltelli, A. (2014). Footprints to nowhere. *Ecological Indicators* 46: 610–621.

41 Goldfinger, S., Wackernagel, M., Galli, A. et al. (2014). Footprint facts and fallacies: a response to Giampietro and Saltelli (2014) "footprints to nowhere". *Ecological Indicators* 46: 622–632.

42 Brundtland, G.H. (1987). *Report of the World Commission on Environment and Development: Our Common Future*. United Nations. World Commission on Environment and Development.

43 Giampietro, M. and Saltelli, A. (2014). Footworking in circles. *Ecological Indicators* 46: 260–263.

44 Van Den Bergh, J. and Grazi, F. (2010). On the policy relevance of ecological footprints. *Environmental Science & Technology* 44 (13): 4843–4844.

45 Stiglitz, J., Sen, A., and Fitoussi, J.-P. (2009). *The Measurement of Economic Performance and Social Progress Revisited. Reflections and Overview Commission on the Measurement of Economic Performance and Social Progress*. OCFE (Centre de reserche en économie de Sciences Po), Paris.

46 Pope, J., Annandale, D., and Morrison-Saunders, A. (2004). Conceptualising sustainability assessment. *Environmental Impact Assessment Review* 24 (6): 595–616.

47 Conrad, Z., Niles, M.T., Neher, D.A. et al. (2018). Relationship between food waste, diet quality, and environmental sustainability. *PLoS One* 13 (4): e0195405.

48 Galli, A., Kitzes, J., Wermer, P. et al. (2007). *An Exploration of the Mathematics behind the Ecological Footprint*. Billerica, MA: Wit Press.

49 Rockström, J., Steffen, W., Noone, K. et al. (2009). Planetary boundaries: exploring the safe operating space for humanity. *Ecology and Society* 14 (2).

50 Certain, G., Skarpaas, O., Bjerke, J.-W. et al. (2011). The nature index: a general framework for synthesizing knowledge on the state of biodiversity. *PLoS One* 6 (4): e18930.

51 Giampietro, M., Mayumi, K., and Ramos-Martin, J. (2009). Multi-scale integrated analysis of societal and ecosystem metabolism (MuSIASEM): theoretical concepts and basic rationale. *Energy* 34 (3): 313–322.

52 Steffen, W., .Richardson, K., Rockstrom, J. et al. (2015). Planetary boundaries: guiding human development on a changing planet. *Science* 347 (6223): 1259855.

53 Jackson, W. and Steingraber, S. (1999). *Protecting Public Health and the Environment: Implementing the Precautionary Principle*. Island Press.

54 Petschel-Held, G., Schellnhuber, H.-J., Bruckner, T. et al. (1999). The tolerable windows approach: theoretical and methodological foundations. *Climatic Change* 41 (3–4): 303–331.

55 Mathias, J.-D., Anderies, J.M., and Janssen, M.A. (2017). On our rapidly shrinking capacity to comply with the planetary boundaries on climate change. *Scientific Reports* 7: 42061.

56 NASA Understanding our planet to benefit humankind: Carbon Dioxide. https://climate.nasa.gov (accessed 16 November 2022).

57 Stockholm Resilience Center. The nine planetary boundaries. https://www.stockholmresilience.org/research/planetary-boundaries/planetary-boundaries/about-the-research/the-nine-planetary-boundaries.html (accessed 16 November 2022).

58 Molina, M.J. and Rowland, F.S. (1974). Stratospheric sink for chlorofluoromethanes: chlorine atom-catalysed destruction of ozone. *Nature* 249 (5460): 810–812.

59 Solomon, S., Garcia, R.R., Rowland, F.S., and Wuebbles, D.J. (1986). On the depletion of Antarctic ozone. *Nature* 321 (6072): 755–758.

60 Stolarski, R.S. (1988). The Antarctic ozone hole. *Scientific American* 258 (1): 30–37.

61 United Nations Environment Programme, Environmental Effects Assessment Panel (2017). Environmental effects of ozone depletion and its interactions with climate change: Progress report, 2016. *Photochemical & Photobiological Sciences* 16 (2): 107–145.

62 Chipperfield, M.P., Dhomse, S.S., Feng, W. et al. (2015). Quantifying the ozone and ultraviolet benefits already achieved by the Montreal protocol. *Nature Communications* 6 (1): 1–8.

63 Ramanathan, V. (2001). Aerosols, climate, and the hydrological cycle. *Science* 294 (5549): 2119–2124.

64 Doney, S.C., Fabry, V.J., Feely, R.A., and Kleypas, J.A. (2009). Ocean acidification: the other CO_2 Problem. *Annual Review of Marine Science* 1 (1): 169–192.

65 Jaramillo, F. and Destouni, G. (2015). Comment on "Planetary boundaries: guiding human development on a changing planet". *Science* 348 (6240): 1217.

Section Five

The Concept of Crisis

11

Collapse and the Anthropocene: Learning from the Past

We are aware that a civilization has the same fragility as a life

Paul Valery (1871–1945)

The collapse of the human population on Easter Island has spawned a vast amount of theorizing. American ecologist, Jared Diamond, argued that this society was undermined by environmental degradation since the seventeenth century. Rapid deforestation along with tensions between rival factions led to a rapid population decline and a loss of sociopolitical complexity.

Image: "Population de l'île de Pâques et statues Moai lors de la visite de l'expédition La Pérouse en 1786. Population of Easter Island and Moai statues during the visit of the La Pérouse expedition" in 1786. Duché de Vancy (artwork: 1797). *Source:* Louis-Antoine Destouff Milet-Mureau / Wikimedia Commons / Public Domain.

A Primer on Human Impacts on the Environment: The Conceptual Approach, First Edition. Liam Heneghan.
© 2023 John Wiley & Sons Ltd. Published 2023 by John Wiley & Sons Ltd.

No entity endures forever; things fall apart. Whenever you observe something that is not in a state of decay, you should ask, why not – how is it resisting deterioration? An interrogation of this sort forms the basis of Erwin Schrödinger's attempt in his short book, *What is Life?* (1944), to define the essence of the living entity [1]. Throughout its lifespan, every living being resists decay up to a certain point at least; the organism retains its order by feeding on order elsewhere. Maintaining itself in a state far from equilibrium is energetically expensive and it requires work. As the organism endures, it struggles against the forces of entropy: it ages and degrades. Ultimately, life cannot and does not contravene fundamental physical laws. When life and its surroundings are considered together, entropy (disorder) always increases – this, as we have seen in a previous chapter, is what forms the indissoluble link between the organism and its environment.

The amount of disorder produced by all human enterprises on a global scale, by means both of our resource use and by the exhaustion of the sinks that absorb waste, is very considerable [2]. The previous chapter considered several frameworks that conceive of a set of planetary limits that can be ruptured by the unimpeded growth of resource extractive economic enterprises. Though each of the models discussed in that chapter that seeks to quantify planetary limits have been subject to legitimate critiques, it is notable, nevertheless, that they all converge in indicating that we are now breaching a number of boundaries in consequential ways.

What exactly is risked by exceeding a planetary boundary? Can breaking these thresholds be potentially catastrophic for humanity [3]? Is it possible that human activity could so fundamentally undermine the stability of the earth's environment that it creates an *extinction risk* for humans? Under what circumstances, then, does it become reasonable to ask if humans, collectively, as an entire species, are in danger of disappearing from the face of the earth?

In what follows, we start by examining the possibility – albeit, an extreme one – that through a radical transformation of the planet upon which we rely, humans might unintentionally undermine our own continued existence. Fears of a human extinction are certainly troubling, though the probability of this eventuality transpiring is considered to be consolingly low, at least in the short term [4]. What of more moderate, though still concerning, speculation that contemporary life is heading into a period of social and demographic decline? To answer this question, environmental writers have often turned to a literature produced primarily by archeologists and anthropologists, which evaluates the circumstances undergirding the unraveling of many past civilizations. Thinking about the historical collapse of civilization might aid in foreseeing what the future may hold for humanity.

The last decades have been ones of unrivaled growth and of increasing economic and social complexity. This is, in part, what is meant by the term the *Great Acceleration*. As more information becomes available about the environmental costs of this untrammeled growth, it seems reasonable to wonder if global civilization can endure in its present form. Is our complex world, which is so energetically expensive to maintain, about to precipitously decline?

11.1 Collapse and the Great Silence

Our universe (and let us assume that you and I inhabit the only universe that exists, despite there being some compelling arguments against this view [5, 6]) is receptive to life in principle – if life occurs here, surely it must occur elsewhere in the imponderable immensity of the cosmos [7]. The contention that there is no a priori reason why life should be restricted to Earth is the foundation for the famous Drake equation. This conceptual equation was proposed by the astrophysicist Frank Drake to estimate the likelihood of intelligent communicative life being found throughout the

Milky Way galaxy [8]. The thinking behind the equation is that there is nothing special about our world: the processes at work on Earth should prevail elsewhere. The assessment that our cosmic situation is a typical one is sometimes called the "principal of mediocrity" – we are neither important nor are we singular. If certain life-supportive conditions line up, then, not only will life emerge, intelligent life should also. The number of intelligent, potentially communicative, civilizations in a galaxy can in theory be estimated. Populating the equation with optimistic values suggests a galaxy that should be teaming with communicators (using highly conservative values in the equation makes the existence of communicators less likely). Despite the plausibility of claims that living beings, eager to communicate with us, must exist is the Milky Way, we still might ask, *"where is everybody?"* This is the question blurted out by physicist Enrico Fermi during a lunch among colleagues at a Los Alamos meeting 1950 convened to discuss the existence of intelligent aliens. It is all very well to conjecture that we live in a region of space teeming with intelligent life, but if this is the case, then why is space so silent? Explaining the *Great Silence* (or *Fermi's Paradox* as the question has also been called) has generated a wide variety of responses. At least one of the proposed solutions to Fermi's paradox is relevant for discussions about the potential extinction of humans on the planet. A comparison of the garrulousness of our planet with the seeming silence of near space reveals two possibilities: either the values often used to populate the Drake equation are wildly optimistic or perhaps every civilization that has evolved elsewhere in the proximate universe had gone extinct!

The **Drake Equation** is usually expressed as:

$$N = R^* \; f_p \; n_e \; f_l \; f_i \; f_c \; L$$

where *N* is the number of intelligent communicating civilizations in the galaxy at present/*R** is the average rate of star formation in our galaxy/*fp* is the fraction of stars that have planetary companions/*ne* is the number of planets per planet-bearing star that have suitable conditions for life/*fl* is the fraction of planets with suitable environments on which life actually starts/*fi* is the fraction of planetary life starts that eventually evolve into intelligent life-forms/*fc* is the fraction of intelligent civilizations that attempt interstellar communication/*L* is the average lifetime (in years) of technically advanced civilizations.

It has proved difficult to populate this equation with confidence. At least some of the attempts in recent years urge us to be optimistic that they are "out there" [8].

Before exploring in more detail these solutions to Fermi's paradox that are relevant for evaluating environmental collapse on Earth, it is worth affirming the conceptual importance of making comparisons between Earth and other worlds. Such comparisons on first inspection can seem fanciful and yet they are an often underappreciated vehicle for thought. The comparative method – which is often employed when experiments or modeling is not possible – is one of the more enduring procedures in the natural and social sciences [9]. The method relies upon the selection of an appropriate *outgroup*, that is, an entity that differs from the one under immediate investigation, but which is the product of the same or related processes. In biology, the method dates back to at least Aristotle (384–322 BCE) who claimed that the functional systems of organisms are best understood by comparing and contrasting among diverse organismal groups. More recently the term refers, more often than not, to a statistical approach that examines correlated traits across a range of species – for example, the relationship between brain weight and body weight in mammals [10]. The comparative method can be adapted to very practical problems in conservation and evolutionary biology [11].

In anthropology, comparisons across culture have been used to test hypotheses about the development of diverse cultural practices. Despite some controversies about its employment in specific circumstances, the validity of the comparative approach, especially when used to pronounce on human universals or to argue about the development of particular aspects of culture, is generally accepted when simple and uniform principles are employed [12, 13]. In employing it, as we are doing here, to discuss the prospects of the collapse of an *entire planetary civilization* (a matter for which no experiment is possible), the comparative method would surely be useful, or at least it would be if we have available to us an appropriate comparative group. In the light of this, Fermi's question, "where is everybody?" is an appeal to the comparative approach and proposes comparing Earth to planets (and satellites) in the Milky Way. If, the same simple and uniform laws apply elsewhere as they do on earth, then, as the Drake equation suggests, the fate of technologically advanced civilization elsewhere (if they can be found) can be usefully compared with Earth. So we ask, why is earth noisy compared to the grave silent of the rest of the galaxy [14]?

The interstellar object "Oumuamua" is most likely a natural object though some astronomers argue that "Oumuamua" is a piece of alien technology. Could it be "a message in a bottle" from an alien civilization?

The object has a range of anomalous properties described by astronomer Abraham Loeb in a paper "6 Strange Facts about the Interstellar Visitor 'Oumuamua'" in Scientific American in 2018 (https://blogs.scientificamerican.com/observations/6-strange-facts-about-the-interstellar-visitor-oumuamua). Loeb writes "An artificial origin offers the startling possibility that we discovered 'a message in a bottle' following years of failed searches for radio signals from alien civilizations."

An artist's impression of "Oumuamua," an interstellar object to pass through the Solar System. *Source:* ESO / Wikimedia Commons / CC BY-SA 4.0.

The *comparative method* (or comparative analysis) is regularly used in the environmental and social sciences to examine how different geographical regions are affected by the same natural environmental processes. Though it is methodological weaker than a well-designed experiment, and often suffers from a lack of replication, it can be helpful in generating hypotheses.

The solutions to Fermi's paradox are both numerous and varied [15, 16]. Some proposed answers are, as we noted above, highly relevant to discussions about the ultimate fate of earthly civilization. Our galaxy is silent, according to one view, because as advanced civilizations approach the complexity where trans-galactic communication becomes possible they accumulate a risk of collapsing. In discussing this viewpoint, philosopher Nick Bostom of Oxford University's Future of

Humanity Institute, building upon the work of economist Robin Hansen, conjectures that the silence is evidence of a barrier – one that he calls a "Great Filter." The Great Filter prevents the emergence of a sophisticated space-faring and communicative civilization because as technological civilization develops, its self-destruction becomes likely [17]. Perhaps merely setting on the path toward the type of technological systems needed for spacefaring, or at least for interstellar communication, sows the seeds of an inevitable extinction. Bostrom writes that "perhaps some very powerful weapons technology" guarantees extinction [17].

More generally, the "sustainability solution" to the Fermi paradox argues for a possibility of collapse of alien civilization through a suite of problems specifically related to environmental mismanagement [18]. Jacob D. Haqq-Misra and Seth D. Baum phrase the sustainability solution in the following way:

> The absence of ETI [Extraterrestrial Intelligence] observation can be explained by the possibility that exponential or other faster-growth is not a sustainable development pattern for intelligent civilizations [18].

Building from this observation, it is at least plausible that our galaxy is silent because any sustainable extraterrestrial civilization must grow very slowly and therefore may not yet have reached across the galaxy. Alternatively, an exponentially growing civilization will have breached planetary limits and will have collapsed. In either of these scenarios, such civilizations would not be detectable to us.

Speculation such as this may seem idle on first inspection but it reminds us, hauntingly, of the inherent fragility of ours being a singular planet: the only one so far with detectable life, the only one with seemingly technologically sophisticated civilization, and one that is at present experiencing vast challenges. In other words, we may be facing into the Great Filter.

11.2 Are the Environmental Sciences Apocalyptic?

Before systematically assessing the potential risks for the thriving of humanity that arise from the breaching of environmental limits, it is useful to evaluate the degree to which environment literature – in both the professional and the more popular forms – is framed by allusions to *apocalyptic* outcomes. From our discussion of the Great Filter, we already have an impression of the contours of the apocalyptic trend in scientific thought. Are conjectures that environmental impacts may result in either the collapse of human civilization or the extinction of humanity warranted? Even if such claims are deemed plausible, is the use of apocalyptic language ever useful? The issue of the apocalyptic gloominess of environmental thought – where appeals are made to potentially terrifying ecological outcomes – is the subject of extensive discussion in the literature of environmental communications, and, more generally, in environmental social sciences (politics and policy discussions, in particular). Reflecting upon this tendency in some environmental writing, the environmental activist Tom Athanasiou speculates in his book *Divided Planet: The Ecology of Rich and Poor* (1998), that some environmentalists have "a professional relationship to their predictions of apocalypse" [19].

What exactly is meant by the apocalypticism of which environmental literature stands accused? How prevalent is it in the professional literature? To what uses are visions of an environmental apocalypse put?

Historically, the term apocalypse was reserved for use in a religious context. In the Christian tradition this is most familiar from the "Revelation" of St. John – the last book of the New Testament, written toward the end of the first century CE, though there are precedents in a variety of older religious traditions [20]. Apocalypticism understood in the sense used in the literature of special revelation, is related to millenarianism, that is, the belief that the end of one age is upon us and a new one about to begin. Despite the prevalence of the meanings associated with religious tradition, The *Oxford English Dictionary* records the following definition of "apocalyptic:"

> Of, relating to, or characteristic of a disaster resulting in drastic, irreversible damage to human society or the environment, esp. on a global scale; cataclysmic.

What is noteworthy is that this definition (which is recorded as a "draft addition" [since March 2008]) *explicitly* invokes disasters that may result in environmental damage. Indeed, the prospect of an apocalyptic future provides an explicit framing for reporting of human impacts on the environment in a variety of newspapers and popular publications. Direct references to an "environmental apocalypse" (or more specifically, a "climate apocalypse" and an "insect apocalypse") are frequently encountered in popular accounts of environmental problems by talented writers. Exemplifying this tendency is an essay by novelist and environmentalist Jonathan Franzen titled "What If We Stopped Pretending? The climate apocalypse is coming. To prepare for it, we need to admit that we can't prevent it" (*New Yorker*, 8 September 2019) in which the word "apocalypse" occurs several times. Franzen writes:

> If you're younger than sixty, you have a good chance of witnessing the radical destabilization of life on earth – massive crop failures, apocalyptic fires, imploding economies, epic flooding, hundreds of millions of refugees fleeing regions made uninhabitable by extreme heat or permanent drought. If you're under thirty, you're all but guaranteed to witness it.

Environmental apocalypticism in media is revealed in the following sample headlines:

Emily Beament, (3 December 2020 Thursday), *Apocalypse Now; Humans Waging "Suicidal War on Nature: UN expert sounds dire warning on climate chaos"* Daily Record and Sunday Mail.

   Damian Carrington (27 November 2020 Friday), *Climate "apocalypse" fears stopping people having children – study.* The Guardian (London).

   Keith Rossiter. (16 November 2019 Saturday), *Is "insect apocalypse" close-by?* Telengana Today. (In the article, Rossiter continues: "Insect apocalypse puts human welfare in peril, experts warn; Decades of abuse come back to haunt humanity as species are pushed to the brink," Keith Rossiter writes).

One other example is instructive: environment editor, Damian Carrington, writes in the *Guardian* newspaper on 27 November 2020 "Climate 'apocalypse' fears stopping people having children – study." Carrington's article reports on a quantitative and qualitative survey of 607 US-Americans between the ages of 27 and 45 evaluating how climate change is factored into their reproductive choice [21]. Almost all respondents (96.5%) expressed that they were "'very' or 'extremely concerned' about the well-being of their existing, expected, or hypothetical children in a climate-changed world." Although the term "apocalyptic" dominated the editorial reflections on this study (and was amplified in extensive social media commentary), the term is rarely found in the original research article that first communicated the survey results. One respondent to the survey wrote, "I feel like I can't in good conscience bring a child into this world and force them to try and survive what may be apocalyptic conditions." The second, and final, use of the term occurred when the authors speculated about the concerns about reproduction in the light of climate change; they suggest it reflects the pessimist views that many Americans have about the future and note the many novels that are now set in "apocalyptic climate futures."

Despite the widespread connection in the popular imagination between anthropogenic damage and apocalyptic outcomes, the use of the term "apocalypse" is rare in the scientific literature. The term is undoubtedly occasionally used in technical literature. A recent example includes a paper titled "Critical warning! Preventing the multidimensional apocalypse on planet Earth" (2020) by Jacques Prieur of the Free University of Berlin in which he calls for a "fundamental societal revolution. . . to preserve the health of populations of living beings, communities and ecosystems." The term "apocalypse," however, only occurs only in the title, though in the paper multiple anthropogenic impacts are considered throughout. More substantive use of the term is found in recent perspective pieces on the reports of global insect declines – however, much of this literature sounds a note of caution about concluding that an "insect apocalypse" is occurring at all [22–24].

In the past five years there have been about 100 published papers in the natural sciences (as revealed by a search of bibliographic website Web of Science) that featured the term "apocalypse" in title, abstract, or keywords. Most of these were concerned with disease (including COVID-19, but including diseases with a longer history in humans), 10 refer to apocalyptic artificial intelligence (including a "robo apocalypse"). Overall, *fewer than 20 papers* discuss an environmental apocalypse (and most of these explicitly reject apocalyptic conclusions). A search of Social Sciences Citation Index and the Arts and Humanities Citation Index yielded 1796 papers employing the term (in striking contrast with the tally from the natural sciences). The majority of these papers are published in Religious Studies journals though as many as 600 are environmentally themed, though it must be acknowledged that many of these papers are an interrogation of apocalypticism and not an endorsement of this way of thinking.

It should be clear from this brief review of the literature that though the prospect of environmentally apocalyptic events looms in the cultural imagination and an investigation of the significance of this phenomenon is taken seriously by scholars, nonetheless the term "apocalypse" is rarely invoked in technical scientific literature on environment problems. Revelations about the apocalypse are largely considered scientifically untenable.

The geographer Erik Swyngedouw concludes his review of environmental apocalypticism as follows:

> In sum, our ecological predicament is sutured by millennial fears, sustained by an apocalyptic rhetoric and representational tactics, and by a series of performative gestures signalling an over-whelming, mind-boggling danger, one that threatens to undermine the very coordinates of our everyday lives and routines, and may shake up the foundations of all we took and take for granted. . . [25]

11.3 Not Apocalyptic but Catastrophic: Environmental Visions of Collapse

There is, quite clearly, an appetite for apocalyptic thinking in some environmental circles. This apocalyptic tendency, as we have seen, is expressed more vehemently in popular writing and environmental journalism than it is in the scholarly literature of the natural sciences. Though explicit invocations of apocalyptic outcomes are seldom encountered in scientific literature, even in that literature there exists quite well developed conjectures arguing for a link between the deterioration of environmental quality and the *demise* of humans – demise usually being a fate somewhat short of total annihilation or apocalypse, but nonetheless catastrophically grim.

One well-considered expression of this concern for humanity's fate is found in *Ecological Literacy: Education and the Transition to a Postmodern World* published in 1992 by environmental studies professor David Orr [26]. Orr's concerns for the future will serve us as a general example of a significant trend in scientifically informed writing on the environment that warns of an impending catastrophe (even as it avoids outright apocalypticism). Orr warns that the world faces three looming crises. Firstly, there is a crisis concerning the provisioning of food for a burgeoning world population (an echoing of the Malthusian crisis). Secondly, we are facing the end of an era of cheap energy (the cultural importance of which is made clear by such writers as anthropologist L.A. White, and historian of technology Lewis Mumford). Finally, there is a crisis related to the breaching of "ecological thresholds" and the limits to natural systems (the topic of the last chapter). Together, these three looming calamities constitute, in Orr's view, the first "planetary crisis." Responses to this planetary crisis would, Orr hoped, "spur humans to a much higher state," or, failing that, Orr dolefully concludes that it will "cause our demise." We have, Orr states, a "decade or two in which to make unpresented changes in the way we relate to each other or to nature."

Orr's account of the crises that humanity faces and the alternatives that confront us is instructive not only because it is elegantly (if sternly expressed), but also because it follows what has become a relatively well-worn pattern in making such assessments. This style of prognosticating the future was not invented by David Orr (though his is a compelling use of it). Firstly, his identification of key environmental challenges is prescient – food security, energy supply, and climate change which remain crux issues today – and his statements also are backed by the best available data on environmental trends. I call this meticulous rostering of all relevant environmental statistics, **data anchoring**. Secondly – and importantly from the perspective of this chapter – Orr pronounces on the stark choices facing us: either a radical change in the conduct of societal affairs, or our imminent demise (**prospect of collapse**). Finally, we are informed that choices for the future must be made very soon (**the narrow timeframe** for action).

Long after Orr's time limit about looming crises was up (he published his warning in 1992 and gave us a decade or so to ascend to a much higher state), we can find his formula out in the (literary) wild. The use of data anchoring, cautioning of the prospect of collapse, and the furnishing of a narrow timeframe – what I call a *tragic triumvirate* – is pervasive.

An example pronouncements about a narrow timeframe for action:

"Only *11 Years Left* to Prevent Irreversible Damage from Climate Change, Speakers Warn during General Assembly High-Level Meeting" (General Assembly "High-Level meeting on Climate and Sustainable Development" [28 March 2019]. For full report see: https://www.ipcc.ch/site/assets/uploads/sites/2/2018/07/SR15_SPM_version_stand_alone_LR.pdf).

We might wonder if Orr was simply incorrect. If indeed Orr was incorrect about the impending planetary crises, he was, at the very least, incorrect in an instructive way. Despite Orr's words of warning, global agricultural systems are certainly continuing to feed a very large population (but with high costs, both economic and environmental). Orr's pronouncements about environmental trends are instructive because the tendencies he noted, with the exception of stratospheric ozone depletion (which is being reversed), have been greatly exacerbated since the early 1990s. Even if there is not agreement about the extent of the crisis we are currently facing, there is a broad acknowledgement by international policy makers that environmental problems are very troubling. We may not have experienced a demise, but we are certainly enduring a global crisis. The *rhetorical effectiveness* of providing a narrow timeframe for action might seem less compelling now that the effects of climate change, biodiversity loss, and soil erosion are now being experienced as omnipresent. According to the models that we considered in the last chapter – especially that of ecological footprinting and the planetary boundary models – we are already bumping up against the limits of the planet. Thus, we cannot forestall environmental deterioration entirely, since that deterioration is manifestly already occurring. There is, however, still time to act, though policy makers seem generally aware that even in acting urgently we cannot avoid damage altogether, rather we act to prevent *the worst implications of environmental change.*

11.4 Definitions at the End of the World

David Orr presents a stark dichotomy: a willingness to address environmental problems may "spur humans to a much higher state" or, ignored, these problems will "cause our demise." However, both options are expressed in nebulous terms. Precisely what is meant by "spurring ourselves to a higher state" is a matter of debate, though, presumably, this must involve transformative change in our sociopolitical arrangements, along with an unprecedented willingness to acknowledge and remediate problems on all scales from the very local to the global [27].

"Demise", the alternative which we may face in the absence of transformative change, is a term that, likewise, is also difficult to pin down. The word derives from the French "démettre" meaning to send away, dismiss, to resign, or abdicate. In common speech, demise equates with death, or cessation of existence, although it is also used as a legal term for the transfer of an estate in a will or a lease [28]. Though the term is used occasionally in the scientific literature that is concerned with the cessation of human existence, it is most often used in relation to the decline of a regional civilization [29–31]. For example, atmospheric scientist A.A. Tsonis and colleagues use the term in describing the role of climate change in the fate of Minoan civilization (an ancient Cretan culture that endured from around 3000 BCE to 1000 BCE). They write:

> Climate change has been implicated in the success and *downfall of several ancient civilizations.* Here we present a synthesis of historical, climatic, and geological evidence that supports the hypothesis that climate change may have been responsible for the *slow demise* of Minoan civilization. [30] (Emphasis mine).

It is clear that "demise" is being used to mean the downfall of that (Minoan) civilization. Brendan M. Buckley and his colleagues use the term demise in a very similar sense in their overview of the contribution of climate to an understanding of the decline of Angkor, Cambodia [29].

Two things seem clear: one, although the term demise is occasionally (though not frequently) used in the technical literature on the (unhappy) fate of civilizations or cultures, it is less prevalent than some of the other terms that we will discuss below. Secondly, when "demise" *is* used, the term is seldom, if ever, defined (though when it is, it usually stops short of an apocalyptic meaning). Rather, the word stands as a vague synonym for other terms like "downfall," which is, itself, a rather imprecise term. This is in sharp conflict with the term "collapse," which is frequently used in the context of social calamities, and suffers, if anything, from a superabundance of precise, though frequently conflicting, definitions. Investigations of the factors contributing to the "collapse" of civilization are so commonplace that we can talk about an emerging discipline of "collapsology" (a term introduced by Pablo Servigne and Raphaël Stevens [32]). Perhaps because the study of collapse coheres into a distinct field, investigators that contribute to these discussions usually state quite exactly what they intend by the use of this word.

> The word **collapsology** (coined by Pablo Servigne and Raphaël Stevens in their book *How Everything can Collapse: a Manual for Our Times* [originally published in French in 2015]) is a umbrella term that brings together all those disciplines that study the collapse of societies (for example, archeologists, anthropologists, environmental writers, catastrophic risk specialists) [32].

Anthropologist and historian Joseph Tainter, the author of *The Collapse of Complex Societies* (1988), a book widely regarded as a classic in this field, cautions about the diversity of definitions of "collapse." Tainter writes:

> Colloquially, collapse means everything from what happened to the Soviet Union to what a worker may do at the end of a hard day. Societies collapse, but then so do bridges, levees, and cardiovascular systems. Academically, the problem is not that definitions vary, but that scholars sometimes discuss the collapses of societies without defining the term [33].

To help with any confusion that the plethora of definitions presents, Tainter offers the following definition of collapse in that book: "A society has collapsed when it displays a rapid, significant loss of an established level of sociopolitical complexity." The Tainter definition, thus, has the following characteristics: collapse is evaluated in terms of (i) a reversal of its sociopolitical structure, and (ii) such a significant reversal occurs over a short period of time.

Jared Diamond, the polymathic ecologist defines "collapse" in a way that resonates with Tainter's definition, though his definition stresses a decline in population as a defining characteristic. In Diamond's best-selling book *Collapse: How Societies Choose to Fail or Succeed*, he writes "By collapse, I mean a drastic decrease in human population size and/or political/economic/social complexity, over a considerable area, for an extended time" [34].

In both Tainter and Diamond's account of collapse the onset of the historical decline in civilization occurs rapidly; some "collapsologists" take exception to Tainter's insistence on the rapidity of this process. Objecting to this insistence on rapid decline, anthropologist Scott Johnson suggests that the term "transition" may be a better descriptor of collapse, since often collapse represents a transition to a different way of life [28].

Johnson writes:

> What archaeologists see as a collapse is usually just a transition to a different way of life . . . the idea of a rapid failure of the systems on which a population depends is intriguing but not an accurate way to describe what happens to most complex societies . . . . "Transition" is a neutral term that better conveys what happens . . . . I use the term "collapse" in a general way, and in most cases I will avoid ambiguity by qualifying what type of breakdown occurred [28].

Mesoamerican archeologist, Ronald Faulseit, in the introductory essay to a book he edited on collapse (*Beyond Collapse: Archaeological Perspectives on Resilience, Revitalization, and Transformation in Complex Societies*, 2016), explores the concepts of collapse, societal transformation, and resilience. He explains that his use of collapse means "the fragmentation or disarticulation of a particular political apparatus," which happens "rapidly" over a few generations [35]. Referring to suggestions about the use of the terms "social change" or "societal transformation" rather than "collapse" Faulseit proposes that societal transformation, his preferred term, can stand as "a broadly defined concept encompassing the full extent of possible outcomes (e.g., collapse, reorganization, revitalization, etc.) associated with societies in transition"; thus "societal transformation serves as an umbrella term that covers the range of sociopolitical trajectories" [35].

---

24   **THE EXAMINER.**

ORIGINAL POETRY.

OZYMANDIAS.

I MET a Traveller from an antique land,
Who said, " Two vast and trunkless legs of stone
Stand in the desart. Near them, on the sand,
Half sunk, a shattered visage lies, whose frown,
And wrinkled lip, and sneer of cold command,
Tell that its sculptor well those passions read,
Which yet survive, stamped on these lifeless things,
The hand that mocked them, and the heart that fed:
And on the pedestal these words appear:
" My name is OZYMANDIAS, King of Kings."
Look on my works ye Mighty, and despair !
No thing beside remains. Round the decay
Of that Colossal Wreck, boundless and bare,
The lone and level sands stretch far away.

                       GLIRASTES.

else be altered ; th
bined, or there s
slating the differer
maintain their own
  No such book i
means get at certa
to be regretted m
Laws ; for in all
tages of one anoth
licitors, who are s
  It often happen
penses of an appc
from the place wh
this parish takes t
which circumstanc
persons never res
by labour or othe
slept a night or tw
shore.
  Two cases occi
above. William

"Ozymandias" by Glirastes, which is a pseudonym for the poet Percy Bysshe Shelley, was published in The Examiner, London, Sunday, 11 January 1818. It is perhaps the best know literary allusion to the collapse of civilization. The poem reflected the growing fascination of the literary public at the time with the discovery of the fragmentary remains of ancient civilization. *Source:* Percy Bysshe Shelley / Wikipedia Commons / Public Domain.

11.5 Ecologizing the End

Instances of the historic "collapse of civilizations" have become the subject of intense academic scrutiny [36]. When environmental scientists conjecture about a potential demise or collapse resulting from an overshoot of resources or from an exhaustion of planetary sinks, they frequently draw upon the work of archeologists and ancient historians [34, 37, 38]. In resorting to the literatures of these other disciplines, do environmental writers accurately report the findings of their colleagues? The answer, generally, is no. Though there are instances where environmental factors clearly underpin a historic collapse – for example, deforestation and severe drought certainly contributed to the collapse in the twelfth century of Chacoan society in northwestern New Mexico, and multiple environmental problems placed strains on the society of Easter Island by the end of the sixteenth century – environmental factors alone are rarely the cause [39].

The literature on "collapsology" is both vast and of very variable quality. This is understandable considering that the topic has captured both the popular imagination and scholarly attention.

Hypotheses about collapse, even in the technical literature, are very often rooted in overly simplified thinking and offer single factor explanations. For example, referring to social conflicts or catastrophic environmental overshoot without reflecting upon interactions between social and ecological factors is probably of limited value. Some explanations rely on supposed mystical forces viewed through the lens of the supposed natural growth and senescence of a culture, though such explanations are usually dismissed by serious scholars of collapse. Moreover, some explanations of collapse apply only over a restricted geographical region or serve to elucidate decline of a small number of societies. Considering the enormous range of examples of collapse, and the various historical times in which the phenomenon has occurred, it is therefore useful to seek conceptual frameworks that can be applied more generally. A general framework is essential if we seek to apply these insights in thinking about the prospect for collapse in our times.

Two accounts, in particular, provide overarching general conceptual frameworks regarding collapse. The first of these is from Joseph Tainter's *The Collapse of Complex Societies* (1988), the other is Jared Diamond's *Collapse: How Societies Choose to Fail or Succeed* (2005). These works are sometimes read as being in opposition to each other. Tainter himself, in remarking on Diamond's work, recognizes that Diamond's framework considers several factors as contributing to past collapses; nonetheless, Tainter claims that several "slips of the pen betray his [Diamond's] conviction that environmental deterioration is really to blame" [39]. That is, Diamond, in Tainter's view at least, finds cases where "bad things happened," and then finds "a plausible environmental reason" for the collapse.

11.5.1 A Five-Point Framework for Collapse

Setting aside Tainter's reservations about Diamond's framework, the framework remains useful at least in this regard: it provides a checklist that can be used to initiate an exploration of any historical collapse. Furthermore, the five-point checklist illuminates past collapses, and it also serves as a tool to investigate strains on our contemporary complex civilization.

The checklist contains the following items: (i) *environmental damage*; (ii) *natural climate change*; (iii) *hostile and warring neighbors*; (iv) the loss of *friendly trade partners*; and, crucially, (v) *society's response to environmental problems*. The sorts of environmental damage that have contributed to collapse include deforestation, habitat destruction, soil problems, issues of water management, overhunting and overfishing, problems with invasive species, especially when these are driven by human population growth and increased per capita environmental impact. Not all five factors apply with equal force in each of Diamond's case studies. The events that transpired on Easter Island prior to European contact in 1722 are the most frequently discussed of Diamond's examples. A brief summary will illustrate how the framework collates disparate factors that can place stress on a society.

There is evidence of extreme *deforestation* on Easter Island (Rapa Nui), prior to 1722, and this contributed to both to soil erosion and the destabilizing of the ecology of the region. Species loss was accelerated by an *invasion* by Polynesian rats (*Rattus exulans*). The remaining trees on the island were *overharvested*, and many food resources were overexploited. Natural *climate change* may have also contributed to the stress on social organization. Though resources on the island became less available, the *population* continued to grow. Warring factions made it difficult for islanders to form a unified approach to these environmental problems [40]. It is worth adding that contact with Europeans after 1722 was, if anything, even more devastating for the people of Rapa Nui than any environmental strains, severe though these must surely have been. Writing in 1994, American anthropologist Douglas W. Owsley remarked, "The eventual result of discovery by Europeans was the near-complete extermination of the population and effective loss of their culture" [41].

The ultimate objective of Diamond's work on collapse was to deduce a framework from the evaluation of historical collapse in order to illuminate what contemporary civilization may be facing in our near future. With this is mind, therefore, we can recognize the destabilizing potential of environmental damage, the severe challenges posed by climate change (this time anthropogenic rather than natural), the annihilative prospects of global warfare, the interdependence of contemporary economies, and the importance of formulating a well-informed response to the issues that confront us.

11.5.2 The Costs and Benefits of Social Complexity: Tainter's Theory of Declining Marginal Returns

In seeking to uncover a range of factors behind the collapse of ancient civilizations, Jared Diamond developed his five-point framework; he determined that some combination of these were always at play when societies became radically less complex (his definition of collapse). His framework was developed as a practical tool for evaluating the fragility of our present moment. In contrast to Diamond, who is primarily an ecologist, the collapse of civilizations is the primary research focus of historian Joseph Tainter's, though he, too, is mindful of the significances of these studies for understanding the risks we face at present.

In his critique of several previous models explaining societal collapse, Tainter argues that these approaches, including Diamond's one, were theoretically inadequate to provide a thorough explanation of collapse. In place of unidimensional explanations of collapse, Tainter proposed an overarching theory based upon four axioms that can then be calibrated to individual circumstances. These axioms build upon each other as follows, (i) human societies are designed to solve problems; (ii) an input of energy is required for the maintenance of sociopolitical systems; (iii) as societies become more complex, the cost of maintenance increases per capita; and, finally, (iv) a point must always be reached where society receives a declining return on its investments (declining marginal returns).

The first two of Tainter's axioms will be familiar to us both from Schrödinger's meditations of the living organism – all ordered entities require energy for their maintenance – and Leslie White's energy and cultural development model. The third and fourth axioms are based upon a well-known law of diminishing returns in economics and other social (and natural) sciences. This law, in general, states that as extra units of energy are committed to production in any system, if all other factors of production are held constant, after a period of increased returns, the resulting output from each additional unit will eventually fall. Inputs committed to the system can be calculated in terms of any energy, labor, or other resources brought to bear on production.

Social collapse can thus be seen as a loss in the capacity needed to maintain social complexity. A loss of social complexity is seen ultimately as an economic matter. For civilizations to be organizationally complex in the first place, they require a constant and positive return on the energetic investment in their maintenance. An environmental or economic stress will always require an organizational response. If the organizational investment made in response to social stress produces a low marginal return, collapse is thus an alternative outcome. To use one of Tainter's examples to illustrate the idea: when the Chacoans (an ancestral Peublo culture centered on Chaco Canyon, between 850 and 1250 BCE) confronted a final severe drought they "collapsed" because doing otherwise would have been too costly compared to the benefits of reinvestment in the social order. Collapse, in this view, is simply "the most logical adjustment" [33].

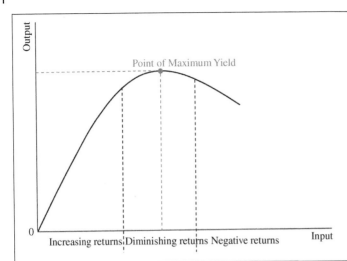

A typical presentation of the law of diminishing returns. Increasing input (of energy, labor, etc.) to the system leads, after a period of increasing returns, to progressively diminishing returns, and subsequently to a negative return. This file is licensed under the Creative Commons Attribution-Share Alike 4.0 International license. *Source:* Author Happy-avocado / Public Domain.

If Tainter's argument is correct, and this principle is a universal one (or at least covering all known instances of collapse), then collapse is predictable based upon analysis of costs and benefits. Any time the cost of a social-organizational change results in a reduced marginal return, then collapse becomes a viable option. Though this model can seem to be reductive, that is, it compresses a diverse phenomenon, that is, collapse, to a single driving factor, and thus becomes guilty of the very flaw that Tainter levies against other theories, he, nevertheless, argues that the theory is more complex than this. The theory of marginal return should be seen as a general overarching principle, and the collapse of individual societies needs to be evaluated in their unique conditions. Local histories are diverse: Tainter's insights need to be sensitively applied.

Tainter's scheme is consistent with arguments about the dangers of overshoot of resources, and the breaching of planetary limits introduced in the last chapter, but it adds a very important dimension to these debates. If the return derived from addressing environmental problems is maintained or increased, then collapse will not happen. However, if the repair of environmental problems is so costly that the expense is not justified then collapse will indeed occur. Therefore, we *cannot* evaluate the risk of collapse under resource stress, as Tainter writes, "without reference to characteristics of the society, most particularly its position on a marginal return curve."

A global catastrophe, such a massive asteroid strike or all out nuclear war, without doubt, could result in a planetary collapse. However, in the absence of an unprecedentedly swift and global blow, collapse that arises from chronic problems including those stemming from environmental deterioration, seem unlikely. This is because the economies of the contemporary world are largely hitched to one another. Some regions are less vigorously linked with the globalized economy than others are but nearly all have partners for both economic and cultural exchange. Under current circumstances, collapse is very unlikely because economic failure within any single region will result in opportunities for expansion for other economies. There are undeniable marginal returns apparent in global affairs with respect to agricultural productivity, mineral and energy production, and the production of new elements contributing to the Gross National Product in industrial countries. The strain that these place on society will be dependent upon the degree to which problems are taken seriously. In this matter, Diamond and Tainter agree: social responses are likely to determine if environmental problems will lead to a collapse. Collapse when it comes will be global and if we allow any collapse at all, disintegration will be total.

11.6 Coda: Human Extinction is Not Inevitable (in the Short Term)

A cottage industry of speculation exists about the moment of first contact between humanity and an alien civilization [42]. The protocols in place to announce such news recognize that there may be consternation in the wake of the news [43]. Depending upon the degree of intelligence of these beings, the implications for science, theology, philosophy, and health could be staggering. If we are contacted (rather than the ones doing the contacting), we might have reason to be concerned since, if human history is a guide, a meetings between cultures typically resulted in the destruction of "contacted" cultures. We have seen with the Easter Island example, but examples of the phenomenon are numerous [44]. So perhaps it should console us that so far space is utterly silent. On the other hand, contact with a technologically sophisticated civilization capable of communication, or even interstellar travel, would suggest that the emergence of such a civilization does not have to result in self-annihilation (that is, it can overcome the Great Filter). Our own planetary society may therefore transcend its problems.

Human extinction, in the short term at least, is not inevitable, but collapse of civilizations, at least those existing on local scales, has been worrisomely frequent in the past. If by collapse we mean a rapid reduction of sociopolitical complexity, accompanied by a precipitous decline in population, then it is collapse, rather than human extinction, that seems like the more pressing concern for present-day humanity. What we have learned from our excursus of the theories of collapse is than no one factor (environmental or otherwise) makes collapse inevitable, but it does become more likely when problems are ignored, and when investments in solutions do not produce benefits valued in excess of the cost of dealing with them. I will leave it to the reader to judge if we are confronting environmental risks with the seriousness that they undoubtedly deserve.

References

1 Schrodinger, E. (2012). *What Is Life?: With Mind and Matter and Autobiographical Sketches.* Cambridge University Press.

2 Nicholas, G.-R. (1971). *The Entropy Law and the Economic Process.* Cambridge, MA/London: Harvard Universiy Press.

3 Bostrom, N. and Cirkovic, M.M. (2011). *Global Catastrophic Risks.* Oxford University Press.

4 Matheny, J.G. (2007). Reducing the risk of human extinction. *Risk Analysis: An International Journal* 27 (5): 1335–1344.

5 Carr, B. and Ellis, G. (2008). Universe or multiverse? *Astronomy & Geophysics* 49 (2): 2.29–2.33.

6 Munitz, M.K. (1951). One universe or many? *Journal of the History of Ideas* 12 (2): 231.

7 Davies, P.C.W. and Davies, P. (2000). *The Fifth Miracle: The Search for the Origin and Meaning of Life.* Simon and Schuster.

8 Angelo, J.A. (2017). Drake equation. In: *Science Encyclopedia: Encyclopedia of Space and Astronomy*, 2e (ed. J.A. Angelo). Facts On File.

9 Sanford, G.M., Lutterschmidt, W.I., and Hutchison, V.H. (2002). The comparative method revisited. *BioScience* 52 (9): 830–836.

10 Felsenstein, J. (1985). Phylogenies and the comparative method. *The American Naturalist* 125 (1): 1–15.

11 Fisher, D.O. and Owens, I.P.F. (2004). The comparative method in conservation biology. *Trends in Ecology & Evolution* 19 (7): 391–398.

12 Mace, R. et al. (1994). The comparative method in anthropology [and comments and reply]. *Current Anthropology* 35 (5): 549–564.

13 Boas, F. (1896). The limitations of the comparative method of anthropology. *Science* 4 (103): 901–908.

14 Tarter, J. (2003). *Ongoing Debate over Cosmic Neighbors*, 46–47. American Association for the Advancement of Science.

15 Cirkovic, M.M. (2018). *The Great Silence: Science and Philosophy of Fermi's Paradox*. Oxford University Press.

16 Webb, S. (2002). *If the Universe Is Teeming with Aliens. . . Where Is Everybody?: Fifty Solutions to the Fermi Paradox and the Problem of Extraterrestrial Life*. Springer Science & Business Media.

17 Bostrom, N. (2008). Where are they? *Technology Review* 111 (3): https://www.technologyreview.com/2008/04/22/220999/where-are-they/.

18 Haqq-Misra, J.D. and Baum, S.D. (2009). The sustainability solution to the Fermi paradox. *Journal of the British Interplanetary Society* 62 (2).

19 Athanasiou, T. (1998). *Divided Planet: The Ecology of Rich and Poor*. University of Georgia Press.

20 Collins, J.J. (2005). Apocalypse: an overview. In: *Encyclopedia of Religion* (ed. L. Jones), 409–414. Detroit, MI: Macmillan Reference.

21 Schneider-Mayerson, M. and Leong, K.L. (2020). Eco-reproductive concerns in the age of climate change. *Climatic Change* 163 (2): 1007–1023.

22 Montgomery, G.A. et al. (2020). Is the insect apocalypse upon us? How to find out. *Biological Conservation* 241: 6.

23 Prieur, J. (2020). Critical warning! Preventing the multidimensional apocalypse on planet earth. *Ecosystem Services* 45: 3.

24 Schowalter, T.D. et al. (2019). Warnings of an "insect apocalypse" are premature. *Frontiers in Ecology and the Environment* 17 (10): 547–547.

25 Swyngedouw, E. (2010). Apocalypse forever? Post-political populism and the spectre of climate change. *Theory, Culture and Society* 27 (2–3): 213–232.

26 Orr, D.W. (1992). *Ecological Literacy: Education and the Transition to a Postmodern World*. Suny Press.

27 Chaffin, B.C. et al. (2016). Transformative environmental governance. *Annual Review of Environment and Resources* 41 (1): 399–423.

28 Dictionary, O.E. *Demise, v*. Oxford University Press.

29 Buckley, B.M. et al. (2010). Climate as a contributing factor in the demise of Angkor, Cambodia. *Proceedings of the National Academy of Sciences* 107 (15): 6748–6752.

30 Tsonis, A.A. et al. (2010). Climate change and the demise of Minoan civilization. *Climate of the Past* 6 (4): 525–530.

31 Yamoah, K.A. et al. (2017). Societal response to monsoonal fluctuations in NE Thailand during the demise of Angkor civilisation. *The Holocene* 27 (10): 1455–1464.

32 Servigne, P. and Stevens., R. (2020). *How Everything Can Collapse: A Manual for our Times*. Wiley.

33 Tainter, J. (1988). *The Collapse of Complex Societies*. Cambridge University Press.

34 Diamond, J. (2011). *Collapse: How Societies Choose to Fail or Succeed*. Penguin.

35 Faulseit, R.K. (2016). *Beyond Collapse: Archaeological Perspectives on Resilience, Revitalization, and Transformation in Complex Societies*. SIU Press.

36 Middleton, G.D. (2012). Nothing lasts forever: environmental discourses on the collapse of past societies. *Journal of Archaeological Research* 20 (3): 257–307.

37 Catton, W.R. (1982). *Overshoot: The Ecological Basis of Revolutionary Change*. University of Illinois Press.

38 Rees, W.E. (2002). Carrying capacity and sustainability: Waking Malthus' ghost. Introduction to Sustainable Development. In: *Encyclopedia of Life Support Systems* (ed. D.V.J. Bell and Y.A. Cheung). UNESCO-EOLSS.

39 Tainter, J.A. (2006). Archaeology of overshoot and collapse. *Annual Review of Anthropology* 35 (1): 59–74.

40 Diamond, J. (2007). Easter island revisited. *Science* 317 (5845): 1692–1694.

41 Owsley, D.W., Gill, G.W., and Ousley, S.D. (1994). Biological effects of European contact on Easter Island. In: *Easter Island in Pacific Context: South Seas Symposium: Proceedings of the Fourth International Conference on Easter Island and East Polynesia* (ed. C.M. Stevenson, G. Lee and F.J. Morin), 129–134. Easter Island Foundation.

42 McKay, C.P. (1936). The search for life in our solar system and the implications for science and society. *Philosophical Transactions of the Royal Society A: Mathematical, Physical and Engineering Sciences* 2011 (369): 594–606.

43 Smith, K.C. and Traphagan, J.W. (2020). First, do nothing: a passive protocol for first contact. *Space Policy* 54: 101389.

44 Crosby, A.W. (2004). *Ecological Imperialism: The Biological Expansion of Europe, 900–1900.* Cambridge University Press.

12

How to Conceive of a (Climate) Crisis

Of course I know in my brain that everything we tell ourselves about human civilization is a lie. But imagine having to find out in real life.

Sally Rooney, *Beautiful World, Where Are You* (2022) [1]

The loss of ice from glaciers around the world is occurring rapidly and accounts for almost 30% of the total observed sea-level rise. Glaciers have already disappeared from many regions of the world and are likely to disappear in all but the most heavily glaciated regions before the end of this century. In addition to contributing to a potentially catastrophic rise is sea level, newly exposed lands differ in the manner in which they reflect solar radiation which, in turn, will affect climate (see, [2]). *Source:* The Art Institute of Chicago / CC0 1.0.

A Primer on Human Impacts on the Environment: The Conceptual Approach, First Edition. Liam Heneghan.
© 2023 John Wiley & Sons Ltd. Published 2023 by John Wiley & Sons Ltd.

We live in a society defined, to a large extent, by its responses to increasingly complex risks: risks emanating from natural causes (for example, earthquakes, volcanoes, and pandemics), risks from warfare and terrorism, and risks emanating from the unintended consequences of our technological way of life – nuclear accidents, chemical spills, and the darker implications of nanotechnology and artificial intelligence – to give just a few examples. That the risks of unintended consequences can be extremely grave are underscored by these remarks by artificial intelligence theorist Eliezer Yudkowsky:

> All else being equal, not many people would prefer to destroy the world. Even faceless corporations, meddling governments, reckless scientists, and other agents of doom, require a world in which to achieve their goals of profit, order, tenure, or other. Therefore I suggest that if the Earth is destroyed, it will probably be by mistake. [3]

Risks from unintended consequences can affect human life either directly or indirectly. For example, the large scale accident in reactor No. 4 at the Chernobyl Nuclear Power Plant on 26 April 1986, resulted in the death of at least 50 people in its immediate aftermath [4]. In addition to the direct effects of this nuclear accident, it also had profound indirect effects by irradiating the environment surrounding the facility: the Chernobyl Exclusion Zone is a 2600 km^2 area of land abandoned (perhaps for future millennia) in the wake of this catastrophe. The longer-term death toll of this accident may end up being close to 4000 people [4–6]. Similarly, the Bhopal disaster on 2–3 December 1984 at the Union Carbide factory, in Bhopal, India, resulted in both an immediate loss of life (deaths are numbered in the thousands) and a suite of long-term health implications; it also necessitated a massive environmental cleanup and rehabilitation of the affected land [7].

The examples of Chernobyl and Bhopal serve to remind us that risk assessors have many variables to consider, the consequences of bad things coming to pass being only one of these. Human well-being in the immediate aftermath of a catastrophic event and the potential for an ongoing human toll needs to be considered, as do the implications for ecological health. Potential disasters vary in the probability of their occurrence, and their assessment must be weighed alongside the (sometimes massive) cost of averting unlikely yet possible events. Awards are rarely handed out for thwarting disasters by means of early action and effective implementation of preventative measures.

In this chapter, we examine approaches to conceptualizing risk; first by illustrating how the assessment of environmental risk has taken its place as a central concern alongside other global threats such as nuclear conflict, hazards associated with pandemic, and catastrophic technological accidents. I describe a model that has been useful in assessing a particular class of risk – global catastrophic risk – and show how it may be used to evaluate environmental risk alongside these other prominent global perils. The chapter ends with a brief assessment of risk from climate change, suggesting that this problem clearly reaches the threshold of being potentially catastrophic in global scope. Climate change may be catastrophic even without considering the possibility that nonlinear surprises in the climatic system may greatly magnify the toll calculated in human welfare and economic costs.

12.1 Assessing Environmental Risk

Many environmental risks can be classified as those arising from unintended consequences. For example, the burning of fossil fuels, which still contributes over three-quarters of the world's energy mix, has vastly increased carbon dioxide concentration in the atmosphere (currently 419 ppm – as of September 2022 – compared with about 278 ppm in pre-industrial times). Alteration of this potent greenhouse gas (one of a number of such trace gases in the atmosphere that act as forcing elements in climate) can modify the regulation of the climatic system in ways that are quite profound (which will be discussed in detail later in this chapter). Likewise, the wide-ranging modification of natural landscapes to accommodate industrial agriculture and urban expansion, as well as the vast appropriation by humans of the products of global ecosystems, including the frequent over-exploitation of individual species to satisfy our wants and needs, has repercussions for the diversity of life on Earth (as discussed in Chapters 13 and 14) [8].

Ulrich Beck (1944–2015), an influential German sociologist, refers to the ways in which we collectively navigate these potential hazards as the "risk society." By this, he means that "we are becoming members of a 'global community of threats'. The threats are no longer the internal affairs of particular countries and a country cannot deal with the threats alone" [9]. Thus, the management of risk has become a global concern. Illustrating the fact that in recent decades environmental risk awareness has taken on greater importance in debates over the future, it is noticeable that Beck's book *Risk Society: Towards a New Modernity* (1992) – an important early analysis of risk society – makes infrequent reference to the political meaning of environmental risk. Yet, in the years that followed, environmental risks, especially those related to climate change, became increasingly central to Beck's work. By the time he wrote and published *World at Risk* in 2009 climate risk was a central aspect of his thinking (to illustrate: there are over 130 references to the climate in that *World at Risk*) [9, 10].

Navigating the rhetoric of risk assessment and appreciating its ramifications in both our personal life (considering the grief, anxiety, and potential apathy that imagining a disaster can induce) and collectively in our political lives (determining priorities with respect to potentially harmful outcomes, planning disaster relief, and managing consequences) is now an inescapable part of an environmental education. Although the recognition of risk associated with global environmental change has been prominent in recent decades, there are now compelling reasons to conjecture that environmental collapse can occur even faster than suspected. Several environment thresholds may be in imminent danger of being breached (as we discussed in Chapter 11): that is, there are boundaries, for example, concerning climate and ecosystem integrity, which, if crossed, will result in modified planetary functioning [11]. These sorts of environmental changes, if they occur, are potentially catastrophic.

In this chapter we will interrogate the concept of catastrophic risk; models of risk that measure the possibility of adverse circumstances affecting human health and mortality, and in large-scale economic losses. Such models allow researchers to compare environmental risks with those that emanate from, for example, natural catastrophes, pandemics, and war. We use estimates of loss, injury, or other adverse effects of climate change as our example throughout this chapter.

12.2 Catastrophic Risk: The Doomsday Clock

In January 2022, *The Bulletin of the Atomic Scientists* re-evaluated the setting of their Doomsday Clock [12]. "Doomsday" in the terminology of this organization is a metaphor for the destructive impacts on our planet stemming from the use of dangerous technologies. The Doomsday Clock identifies those perils we must be confronted if humanity is to endure. The clock remained at 100 seconds to midnight in 2022, the same as it had been since 2020 when the clock had been moved forward from two minutes before midnight. The Earth in the assessment of the Bulletin's expert panel remains in grave peril. *Three factors* in particular are identified as the drivers of this immanent doomsday.

1) *The global pandemic* of COVID-19 declared by the World Health Organization in March 2020, which as of writing this in November 2022 has claimed almost 6 602 000 million lives [13].
2) *The prospect of nuclear annihilation.* This risk factor provided the original motivation for the clock when it was first set in 1947. At that time, the prospect of a proliferation of nuclear weapons fueled a fear of a nuclear arms race. Fears about the deployment of nuclear weapons have become widespread in 2022 accompanying the conflict in Ukraine.
3) A third factor – one that has only been included in the roster of doomsday vectors only since 2007 – is the prospect of *catastrophic disruptions from climate change.*

The Doomsday Clock is based upon a risk assessment exercise performed annually by an interdisciplinary expert panel. The inclusion of climate risk as a factor in setting the clock reflects the growing concern over environmental problems (a concern that we noted as being also reflected in the work of Ulrich Beck, a leading risk theorist). The clock is ultimately an advocacy tool; it seeks to elevate the levels of concern paid to catastrophic risk with the hope that disaster can be averted. Important from the perspective of our discussion in this book is the fact that the doomsday analysis puts into conversation discourses about sets of individual risks (nuclear, disruptive technologies, pandemics, and environmental disruptions) that had formerly been held separate. The fact that these risks are being considered together is important and appropriate, as international cooperation is required to reduce both nuclear and technology threats as well as climate change. Since the Doomsday Clock was developed and is currently maintained by a highly interdisciplinary panel, the criteria upon which the decisions are based are, by necessity, *subjective.* Below we ask if we can designate a set of *objective* criteria that can be employed to make side-by-side comparisons of the full panoply of catastrophic risks. It is worth noting, however, even objective criteria must be sensitively applied, and this application is itself subjective.

12.3 Catastrophic Risk Assessment

There is no universally recognized definition of what constitutes a catastrophic risk [3]. In general, any eventuality that has widespread implications for human well-being, or that can cause a substantial increase of mortality on local, regional, and global scales can be regarded as a catastrophe. This definition is in some ways vague but it is clear in at least one important respect: it places the emphasis on the *outcome* of the catastrophic event rather than the cause. A fire, a disease outbreak, or an environmental problem is locally or regionally catastrophic to the degree that it affects human welfare, and not merely because the event has occurred in the first place. In much of the contemporary literature on catastrophic risk, outcomes that are considered to be useful metrics include not only human causalities or fatalities but also include economic losses. According to philosopher Nick Bostrom and astronomer Milan M. Ćirković, who jointly edited an influential volume on a vast range of Global Catastrophic risks (*Global Catastrophic Risks.* Oxford University Press, 2011), an event that resulted in 10 000 000 deaths and $10 trillion in economic loss would generally constitute a global catastrophe, whereas an event resulting in one million dead and one trillion dollars in economic harm would not rise to that threshold. However, they recognize that no clear quantitative threshold exists that could be universally applied in making a determination about a catastrophe. The COVID-19 pandemic has resulted in 6 602 000 deaths (as we noted above) and at least $12 trillion in economic cost from 2020 to 2021. If this inarguably horrific loss of life does not quite meet the toll that Bostrom and Ćirković proposed in their loose definition, it should be clear that this is because at least some of the large economics costs associated with the pandemic were a consequence of the strenuous and costly efforts to minimize the death toll. Few would dispute that COVID-19 was a global catastrophe.

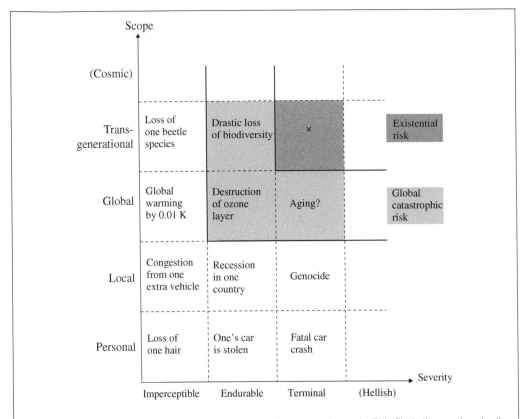

A qualitative categorization of different risks in terms of scope and severity. This file is licensed under the Creative Commons Attribution-Share Alike 3.0 Unported license. *Source:* Author Wrev / Public Domain.

Bostrom and Ćirković propose a helpful model to evaluate risks emanating from both natural and non-natural sources. Risk levels can be distinguished by their *scope* (how many people would be affected), its *intensity* (how badly these would be affected), and the *probability of its occurrence*. The term scope also records the geographical extent of the fallout from the impact. The intensity with which we experience *events* in the world can range from *imperceptible* to *terminal*. By terminal is meant death, though the editors propose an additional category, that is, the category of *hellish*, one that acknowledges that there are things in this universe that are worse than death. To clarify this scale of intensity: most of us would consider that the risk emanating from a hair falling from one's head is personal in scale and imperceptible in scope, whereas a car accident, while also personal, may be terminal.

Of particular concern for conducting risk analysis are those risks that cause loss of life and are global in scale. Such risks are classified as *global catastrophic risks*. What sorts of events result in devastation on this scale? These risky events can arise from *natural causes* or from *anthropogenic ones*. The former category includes asteroid collisions, super-volcanoes, dramatic (natural) climate change, or indeed the eventual engulfing of the planet by the expanding sun, and the culminating heat death of the planet. Anthropogenically caused catastrophes include a variety of factors but frequently include the accidental implications of technology gone awry. Nuclear accidents, biotechnology misadventures, and unfriendly AI all serve as examples. Some of the implications of technology can be intentional, of course. Though it seems unlikely that anyone would intentionally unleash a catastrophe that engulfs the entire globe, there is quite obviously a substantial investment in technologies developed for military purposes that could do precisely this (especially when such systems are linked into tit-for-tat guarantees of reprisals, such as nuclear mutually assured destruction). In the event of a global conflict, these can produce outcomes that inflict harm on a very vast geographical scale. It was reflecting on these sorts of catastrophes that led to the creation of the Doomsday Clock in the first place [12].

Many of the potentially catastrophic events mentioned above have prompted efforts to avert with the risk. For example, a program to detect earth-approaching asteroids is in place. Other conceivable threats are regarded as more theoretical (and little or no preventative infrastructure in put in place); such threats include one from high-energy physics experiments where the risks seem remote [14]. Determining which risks should be regarded as a priority and which might be relegated to a category of lesser concern, are the subject of comprehensive risk analysis.

Risk analysis consists of the following steps: a characterization of risk, an estimation of its probability and, finally, an evaluation of likely consequences. The characterization of risk requires meticulous enquiry about the scientific underpinnings of the causes and consequences. Those events that are most likely to have globally catastrophic consequences are those that are both infrequent and improbable. Volcanic eruptions, for example, are commonplace, but most have few consequences for humans; some volcanoes, however, have local impacts, and super volcanic eruptions are exceedingly rare and have devastating implications. For example, the Toba volcano, which occurred between 60 000 and 75 000 years ago, is estimated to have been 3 orders of magnitude larger than the 1980 eruption of Mount St. Helens and may have caused a global cooling that persisted for at least several decades (if not longer). As a result, the human population, which was globally rather small at the time, may have been drastically reduced. Though the claims are controversial, the drop in human numbers may have resulted in a genetic bottleneck for the human population. This reduction of population was likely as a consequence of the reduction of plant productivity caused by dust in the atmosphere. The example of the Toba volcano is interesting (though anxiety producing), as it underscores the possibility that a rare but catastrophic event can bring humanity close to extinction.

The frequency distribution of risk is important to consider. Global catastrophic risk tends to be fat-tailed risk, meaning that in several categories they have a larger probability of occurring than would otherwise be anticipated. Though such risks can be characterized when the scientific data are available, however, policy making can be stymied as a result of some cognitive biases surrounding these sorts of risks: we tend to underestimate the occurrence and the implications of rare events no matter how well categorized they are, especially when they seldom occur in the lifetime of the individual making the assessment.

12.4 Climate Change: A Brief Primer

In what follows I provide a short sketch of the causes and consequences of climate change. Climate change was one of nine planetary boundaries that we considered in Chapter 11. This is decidedly not a comprehensive treatise on the topic, but is merely offered as a case study to address questions concerning the risk associated with global environmental problems. Do such risks rise to the level of global catastrophic risk, a category of risk introduced above?

12.4.1 Causes of Climate Change

In the same year Charles Darwin published his epoch-making book *On the Origin of Species*, the Irish physicist John Tyndall (1820–1893) initiated his groundbreaking work on infrared radiation and on functional aspects of the atmosphere. Tyndall's influential paper "On the Absorption and Radiation of Heat by Gases and Vapours, and on the Physical Connexion of Radiation, Absorption, and Conduction" was published in 1861 and reported on a series of experiments into how thermal radiation behaves in its interaction with atmospheric gases [15]. Using an instrument called a ratio photospectrometer, Tyndall observed that some gases, for example carbon dioxide and water vapor ("aqueous vapor" as he called it), help stabilize the temperature of the earth. Changes in these gases, he wrote, sagely, "may in fact have produced all the mutations of climate which the researches of geologists reveal."

Though we should caution ourselves against reading back through the sequence of scientific discoveries in the light of current concerns, nevertheless, Tyndall's pioneering research has taken on a magnified significance in the century after he undertook it in the light of apprehensions about anthropogenic climate change. Tyndall undertook his work in the context of debates at the time about thermodynamics – his main concerns were with concepts of heat. This work on atmospheric gases is particularly significance since his observations were critical in the evolution of our understanding of the planet's heat budget.

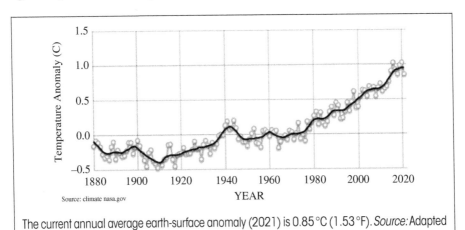

The current annual average earth-surface anomaly (2021) is 0.85 °C (1.53 °F). *Source:* Adapted from Global surface temperature relative to 1951–1980. (Source nasa.gov, public domain).

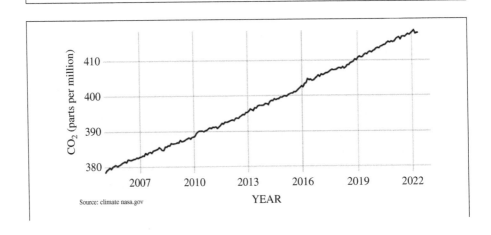

Carbon dioxide content: 418 ppm (May 2022). Human activities have raised atmospheric concentrations of CO_2 by about 50% since preindustrial time.

NASA content - images, audio, video, and computer files used in the rendition of three-dimensional models, such as texture maps and polygon data in any format - generally are not subject to copyright in the United States. You may use this material for educational or informational purposes, including photo collections, textbooks, public exhibits, computer graphical simulations and Internet Web pages. This general permission extends to personal Web pages.

Tyndall's work contributed to our understanding of the Earth's heat budget (also called, the Earth Radiation Budget ERB). The ERB accounts for the input and output of solar energy at a global scale. The balance of input and output of energy determines the average global average temperature. Simply stated, the Earth receives energy from the sun in the form of short wave radiation. The total radiation (the "solar constant") from the sun has been measured as $1366\,W\,m^{-2}$, with some variation due to solar variability. Some of this energy is radiated directly back from earth (a fraction that is called albedo). Brighter areas, like snow pack, reradiate a larger fraction of energy back to space than darker areas that absorb more. The net incoming energy is greatest at the equator, and is reduced poleward. In the absence of the atmosphere, calculating the ERB would be a relatively uncomplicated affair. Temperatures on earth in such circumstances would be well below zero, approximately 33°C cooler than they are today, and human life could not be sustained [16]. Tyndall's pivotal work revealed that greenhouse gases in trace amounts are substantial contributors to the ERB.

The Earth radiates energy back into space mainly in the form of long-wave, black body radiation. Averaged over the course of a year, the incoming radiation is balanced by outgoing radiation. This maintains the heat balance of the Earth. Since there is more solar radiation incoming to the tropics and more outgoing from the poles, there is, as a consequence, a considerable transportation of energy from tropics toward the poles – a transfer of energy that drives important climatic patterns. The transfer involves both atmospheric as well as oceanic transfer.

In the work of the IPCC, *radiative forcing* is defined as "the change in net downward radiative flux at the tropopause [the lower atmosphere] after allowing for stratospheric temperatures to readjust to radiative equilibrium" [17].

To understand the regulatory importance of greenhouse gases (and, indeed to fully appreciate the challenges of humanmade climate change) is to distinguish between climate "forcings" and "feedbacks." *Radiative forcing* is quite simply the change in the rate of energy received on Earth through solar radiation and its rate of dissipation (once again, this is measured in Watts per square meter [$W\,m^{-2}$]); when radiative forcing is positive the Earth receives more incoming energy from sunlight than it radiates to space and will result in warming. Evaluating changes in radiative forcing is key to identifying the human "fingerprint" on climate regimes. Forcing includes a variety of factors, including changes in the sun's energy output, variations in the Earth's orbital cycle, the contribution of volcanic eruptions, and radiatively active greenhouse gases.

Greenhouse gases affect radiative forcing. These gases, as Tyndall discovered, absorb *infrared* radiation. Once absorbed, infrared radiation must be re-emitted. The re-emission from greenhouse gases travels in all directions. Roughly, half the radiation is emitted downward and half upward [16].

The back radiation heats the lower atmosphere. For this reason, the heat that we experience on the surface of the earth comes from both the sun and the warming effect of greenhouse gases. According to climate scientist Kerry Emanuel – whose excellent short account of contemporary climate change *What we know about climate change* (2018) I recommend – the planetary surface receives about twice as much radiation from the atmosphere as it does directly from the Sun [18].

A climate feedback according to the Intergovernmental Panel on Climate Change (IPCC) represents the "interaction mechanism between processes in the climate system... when the result of an initial process triggers changes in a second process that in turn influences the initial one. A positive feedback intensifies the original process, and a negative feedback reduces it" [17].

In addition to understanding the concept of radiative forcing, the importance of *climate feedbacks* need to be appreciated. A feedback is a process that reacts to ongoing changes in the climate itself. For example, as the global temperature rises, it increases the amount of water vapor in the atmosphere (through evaporation); since water vapor is a potent greenhouse gas this interaction results in even more heating. This would be considered a positive (potentially destabilizing feedback).

12.4.2 The Human Manipulation of Climate

Water vapor, a greenhouse gas, can be present in the atmosphere in relatively high concentration. The presence of water in the atmosphere as clouds can also have striking implications for climate. However, it is by means of the large-scale inadvertent experiment of the carbon cycle by humans that we most affect climatic systems. Carbon dioxide has, after water vapor, the largest effect on surface temperature. Methane and nitrous oxide are both more potent greenhouse gases than carbon dioxide. However, though the emissions of both are influenced by humans, these emissions are in smaller amounts than carbon dioxide.

The releasing of vast amounts of carbon into the atmosphere – the large-scale experiment I just alluded to – in the form of carbon dioxide began in earnest in the industrial revolution. Fossil fuels, in the form of coal, oil, shale, and natural gas, provided the energy needed for revolutionary change in human production of goods. These resources were formed during the carboniferous and tertiary geological periods when the remains of plants and microbes were incompletely decomposed (thus retaining large amounts of unexploited energy). Since the beginning of the industrial era, human industrial and domestic endeavors have essentially reversed a process of carbon accumulation that took a geological epoch to complete. The fate of this material was always going to be realized through oxidation, but humans have acquainted this carbon with its fate by burning vast quantities of it in a geological instant, and thereby adding huge amounts of carbon dioxide to the atmosphere. More than three-quarters of carbon dioxide emissions come from this combustion of fossil fuels.

That this massive release of carbon dioxide would have consequences for the planetary heat budget was realized soon after the discharge began in earnest. The Swedish chemist and Nobel laureate Svante Arrhenius published a scientific paper in 1896 and later a book (1908) that predicted that a doubling of carbon dioxide in the atmosphere would elevate surface temperature of the planet by about 4°C (his early estimates were higher). This prediction is not wildly inconsistent with contemporary the predictions of the IPCC [19].

In a previous chapter, we discussed the work of Charles Keeling, whose pioneering work on atmospheric carbon monitoring, alerted the global scientific community to the ongoing elevation of carbon dioxide, which now stands at 419 ppm [20]. Keeling's work that highlighted the degree to which contemporary levels of carbon dioxide exceed preindustrial levels (278 ppm) ignited renewed interest into the question of how human modification of the carbon cycle will influence the stability of climate systems. In the later part of the twentieth century, climate science has undoubtedly intensified. It has become a vastly interdisciplinary endeavor seeking to quantify the causes and consequences of these recent changes to global functioning. In addition, the climate discipline assess the social implications, the formulation of appropriate policy responses, and the effectiveness of the implementation of these strategies [21].

12.5 Climate Change and Catastrophic Risk

12.5.1 Ongoing Changes Are Severe...

A range of potential consequences derives from the manipulation of greenhouse gas levels in the atmosphere. Among the many effects the following are especially significant.

1) Since the late nineteenth century, the estimate for the amount of global warming that has occurred is 1.07 °C (compared to the average from the decades before the industrial period). The main driver of warming in the lower atmosphere has been the anthropogenic input of greenhouse gases. This level of warming is unprecedented in the last 2000 years if not longer, and the carbon dioxide level in the atmosphere is higher than at any other in at least two million years.

2) Climate change has led to an intensification of the water cycle: more intense rainfall, elevated flooding, as well as more intense droughts. In the higher latitudes, precipitation is likely to increase, though decreases will occur in the subtropics with grave ecological consequences. Changes to monsoon precipitation are ongoing and more are expected – these rains are vital to the economies of monsoonal regions.

3) During the twentieth century and on into the early twenty-first century, sea level has risen faster than over any preceding century in at least the last 3000 years. Additionally, the ocean has warmed in this past century more that it has since the end of the last Ice Age (approximately 11 000 years ago). Open ocean pH has decreased since the beginning of the industrial era. Sea level will continue to rise throughout the twenty-first century, resulting in severe coastal flooding and intensifying coastal erosion. The occurrence of coastal flooding events will become more frequent.

4) Permafrost thawing will occur more rapidly; additionally, there is ongoing loss of seasonal snow cover in various regions, an acceleration in the melting of glaciers in addition to the retreat of ice sheets. Summer Arctic sea ice is expected to disappear, (being in the last decade at the lowest level it's been since the 1950s).

5) More intense heat in many cities is expected.

The combination of changes in temperature, alterations of the hydrological effects, sea level, loss of permafrost, and intensified heat island effect, are all consequential in themselves. However the indirect effects on agricultural productively, on changes to the behavior of disease vectors, the implications for biodiversity loss, and the consequent implications for human health, wellness, and habitation, intensify all of these effects.

12.5.2 ...but Are These Effects Potentially Catastrophic?

The IPCC projections of future climate change – that is, warming somewhere between 1.4 and 5.8 °C (in somewhat extreme projections) by 2100 – are worrisome enough without us having to invoke nonlinear calamitous change. There has been an understandable reluctance of scientist to invoke the prospect of immense catastrophes, erring, as we have previously discussed, on the side of "least drama" [3, 22].

There is, however, no doubt that self-reinforcing, positive, feedbacks could nudge Earth systems toward thresholds that, if breached, would make stabilization of the climate very difficult to achieve. In such circumstances human contributions to climate change become moot, because feedbacks kick in that seek a new equilibrium temperature for the planet [23]. In such circumstances, warming may accelerate and then become harder to stop [24]. Scientists are reluctant to evoke such scenarios for a variety of reasons, one of which is that it becomes impossible to predict the future and science loses it regulatory input into policy decisions. However, several credible scenarios exist for runaway climate change even though each has a relatively low probability of happening.

1) *Methane hydrate risk*: large quantities of methane (CH_4) can potentially be released into the atmosphere by means of the destabilization of hydrate deposits trapped in marine sediments. This destabilization can occur as a result of warming of both ocean waters and deep marine sediment. Methane is an extremely potent greenhouse gas (25 times more potent that carbon dioxide) and massive release from marine sediment could be a tipping point for additional climate change [25].
2) *Methane from permafrost*: a credible additional source of methane (CH_4) is the thawing of permafrost released because of enhanced decomposition of dead plant matter in these soils. Estimates of the likelihood of billions of tons of CO_2 and CH_4 emanating from this source are not fully developed but the potential impacts are so high that urgent action to minimize the scale of the release should be undertaken [26].
3) *Thermohaline disruption*: a slowdown or indeed a shutdown of the thermohaline circulation would have dramatic implication for climate change. The dramatic effect of a slowdown (which is perhaps already underway) does not, in fact, accelerate warming, rather it would promote global cooling [27].
4) *Iris effect*: though the effect is not well quantified, the analyses of water vapor and clouds in the tropics suggest the emergence of an "infrared iris" that may regulate the heat budget in these regions. If such a mechanism is activated it would reduce the impacts of warming in tropical regions [28].
5) *Cloud feedback*: the effect of cloud feedback on climate change is complex and there are many uncertainties about these mechanisms [29]. In some scenarios, feedback from cloud changes amplifies the effects of climate change.

So, is climate change a potential global catastrophic risk? If you are reading these pages in the shadow of the global COVID-19 pandemic, you have an intuitive understanding of what that rapidly unfolding catastrophic event entailed. Lockdowns, a massive economic investment both in research and in ameliorative steps, and navigating high large levels of uncertainty, all are occurring as infections escalate and the death toll rises (in excess of a half a billion case and more than 6 602 000 deaths). This pandemic will remain exemplary for the present generation as a paradigmatic global catastrophe. Climate change might never be quite so immediately conspicuous since it is unfolding over the course of many generations and will be somewhat patchy in it geographical effects. If our criterion for a global catastrophe include a large geographic extent while still being endurable despite increased mortality and extensive economic harm, then undoubtedly climate has already reached this threshold. If we insist on quantifying the toll (say, 10 000 000 deaths and $10 trillion in damage) then it remains harder to say, since the tools for quantifying deaths and economic harm due to climate change are in their infancy. The World Health Organization estimates that between 2030 and 2050 climate change will cause 250 000 additional deaths, and that the health costs alone are estimated to be US$ 2–4 billion/year by 2030.

It is eminently reasonable, based upon a recognition about the consequences of breaching important thresholds to be concerned that the effects of climate change, considered in the context of the entire global social-ecological systems, may be catastrophic.

12.6 Coda: A Note on Eco Anxiety

Although a full accounting of human impacts on the environment should encompass the helpful aspects of environmental management, while providing an account of appropriate technologies and policy interventions beneficial for both humans and the rest of nature, nevertheless, the focus in *this book* (and the course upon which it is based), is, rather, on understanding the scope and consequences of environmental problems. But not all environmental problems produce catastrophic outcomes. Indeed not all environmental systems are equally fragile: everything that humans touch does not necessarily fall apart. It is just as much a mistake to assume no vulnerability (in the absence of evidence) as it is to assume that every system is equally (and highly) vulnerable to environmental damage.

However, in concentrating on the issue of limits and on the risks to both human and environmental health, the approach we are taking here is not atypical. For better or worse, a discussion of environmental solutions is left to other books (and classes). However, scientists and policy makers (and indeed writers of books) alike needs to be attentive to the ways in which risk evaluation is presented. Oftentimes, the problems that we discussed – climate change, biodiversity loss, population issues, and so forth – are presented as potentially catastrophic in their implications. Concerns about catastrophic perils can serve to focus attention on issues of the utmost priority – it's hardly surprising that warnings about human impacts can seem grave.

Catastrophizing situations – that is, treating possibly innocuous situations as being considerably worse than they may turn out to be – can, however, have the counterintuitive effect of diminishing an appetite for engaging with problems [24]. Seeming apathy in the face of environmental catastrophe must be interpreted with caution. What can seem like apathy should not necessarily be regarded as a lack of concern but, rather, can be seem, according to environmental psychoanalyst Renee Aron Lertzman, as expressions of "difficult and conflicting" emotions [24]. In facing a disaster, it can be hard for many people to see how they can contribute to the solution. For this reason, environmental scientists and advocates must be ever mindful of the ways in which the tone used to communicate the environmental situation, can generate anxiety and guilt. It is, for this reason, valuable to assess questions of resource management and policy interventions in the context of system resilience (defined here as the capacity of a system to absorb an impact without a fundamental alteration of the structure or function of a system) [30].

A consequence of an approach that focuses on the negative outcomes of environmental impact can be personally distressing and may produce high levels of despair and anxiety in students of the environment (as well as for attentive readers of emerging environmental research and literature).

Recognizing this, there is now a growing attention among environmental psychologists and social scientists to the full dimensions of *eco-anxiety* or related concerns about the trauma of "ecological grief." By *eco-anxiety* is meant "the anxiety related to current and predicated environmental damage or loss, particularly from the climate crisis," and *ecological grief* is "the grief and sadness felt in response to the loss of beloved places, ecosystems, and species" [31]. Anxious reflection about the future can, in some circumstances, be highly motivating and can potentially increase environmental engagement. However, though it is tempting to try to short-circuit anxiety concerning environmental impacts by committing to immediate action – an approach summarized by the adage: "the antidote to anxiety is action" – *unreflecting* action is not always helpful, and can often be economically costly and environmentally damaging in its own way. This is not only because a lack of due consideration can result in underestimating the magnitude of our problems and, oftentimes, results in a failure to identify the most relevant forms of actions [32]. Recent research indicates that increased engagement in environmental projects is a better salve for anxiety when this activity is directed at real solutions [31]. Therefore, pausing to reflect and to acquire knowledge relevant for the most effective environmental solutions is vital. To put this another way: reflection is a form of action.

Though it may be troubling to consider, and many environmental thinkers counsel us to remain hopeful about the future, we cannot, at the same time, falsely minimize the immensity of our challenges. An honest reckoning of the issues remain paramount. Such an honest approach seem wise considering the degree of attention in recent years being paid in scientific journals to the risk of catastrophic surprises. Consider, for example, a paper published in the influential journal *Nature Climate Change* in 2014, where Christian Franzke, a physicist working at the IBS Center for Climate Physics (ICCP) wrote that "we urgently need to better understand the natural decadal scale and the nonlinear aspects of the climate system" [33]. But Franzke's warning was just one example of those coming from several scientists who were prepared to study potentially catastrophic surprises in climatic and ecological systems (for examples of these see [34–36]). These reflections are often concerned with "tipping points" in the climatic system but such thresholds exists is a wide variety of ecological systems: lakes, oceans, forests as well as in human societies [37]. We will need, therefore, to balance fierce hope with a relentlessly truthful assessment of the future.

References

1 Sally, R. (2022). *Beautiful World, Where Are You*. Faber & Faber.

2 Zemp, M. et al. (2019). Global glacier mass changes and their contributions to sea-level rise from 1961 to 2016. *Nature* 568 (7752): 382–386.

3 Bostrom, N. and Cirkovic, M.M. (2011). *Global Catastrophic Risks*. Oxford University Press.

4 Bohannon, J. (2005). Nuclear medicine – panel puts eventual Chornobyl death toll in thousands. *Science* 309 (5741): 1663.

5 Baverstock, K. et al. (1992). Thyroid cancer after Chernobyl. *Nature* 359 (6390): 21–22.

6 Steinhauser, G., Brandl, A., and Johnson, T.E. (2014). Comparison of the Chernobyl and Fukushima nuclear accidents: a review of the environmental impacts. *Science of the Total Environment* 470: 800–817.

7 James, S.L. et al. (2018). Global, regional, and national incidence, prevalence, and years lived with disability for 354 diseases and injuries for 195 countries and territories, 1990-2017: a systematic analysis for the Global Burden of Disease Study 2017. *Lancet* 392 (10159): 1789–1858.

8 Imhoff, M.L. et al. (2004). Global patterns in human consumption of net primary production. *Nature* 429 (6994): 870–873.

9 Beck, U. (2009). *World at Risk*. Polity.

10 Beck, U. (1992). *Risk Society: Towards a New Modernity*. Sage.

11 Steffen, W. et al. (2015). Planetary boundaries: guiding human development on a changing planet. *Science* 347 (6223): 1259855.

12 Science and Security Board of the Bulletin of the Atomic Scientists (2021). Doomsday Clock Statement. *Bulletin of the Atomic Scientists* https://thebulletin.org/doomsday-clock/current-time.

13 Coronavirus Cases (2022) https://www.worldometers.info/coronavirus (accessed 16 November 2022).

14 Koch, B., Bleicher, M., and Stöcker, H. (2009). Exclusion of black hole disaster scenarios at the LHC. *Physics Letters B* 672 (1): 71–76.

15 XXIII Tyndall, J. (1861). On the absorption and radiation of heat by gases and vapours, and on the physical connexion of radiation, absorption, and conduction.—the bakerian lecture. *The London, Edinburgh, and Dublin Philosophical Magazine and Journal of Science* 22 (146): 169–194.

16 Emanuel, K. (2018). *What we Know About Climate Change*. MIT Press.

17 Pörtner, H.-O. et al. (ed.) (2022). *Climate Change 2022: Impacts, Adaptation, and Vulnerability. Contribution of Working Group II to the Sixth Assessment Report of the Intergovernmental Panel on Climate Change*. Cambridge University Press.

18 Soares, M.D. et al. (2020). Oil spill in South Atlantic (Brazil): environmental and governmental disaster. *Marine Policy* 115: 7.

19 Rodhe, H., Charlson, R., and Crawford, E. (1997). Svante Arrhenius and the greenhouse effect. *Ambio* 26 (1): 2–5.

20 NASA Carbon Dioxide. https://climate.nasa.gov (accessed 16 November 2022).

21 Shaman, J. et al. (2013). Fostering advances in interdisciplinary climate science. *Proceedings of the National Academy of Sciences* 110 (supplement_1): 3653–3656.

22 Brysse, K. et al. (2013). Climate change prediction: erring on the side of least drama? *Global Environmental Change* 23 (1): 327–337.

23 Steffen, W. et al. (2018). Trajectories of the earth system in the Anthropocene. *Proceedings of the National Academy of Sciences* 115 (33): 8252–8259.

24 Lertzman, R.A. (2012). The myth of apathy: psychoanalytic explorations of environmental subjectivity. In: *Engaging with Climate Change* (ed. S. Weintrobe), 139–165. Routledge.

25 Ruppel, C.D. (2011). Methane hydrates and contemporary climate change. *Nature Eduction Knowledge* 2 (12): 12.

26 Hope, C. and Schaefer, K. (2016). Economic impacts of carbon dioxide and methane released from thawing permafrost. *Nature Climate Change* 6 (1): 56–59.

27 Anthoff, D., Estrada, F., and Tol, R.S.J. (2016). Shutting down the thermohaline circulation. *American Economic Review* 106 (5): 602–606.

28 Lindzen, R.S., Chou, M.-D., and Hou, A.Y. (2001). Does the earth have an adaptive infrared iris? *Bulletin of the American Meteorological Society* 82 (3): 417–432.

29 Gettelman, A. and Sherwood, S.C. (2016). Processes responsible for cloud feedback. *Current Climate Change Reports* 2 (4): 179–189.

30 Folke, C. et al. (2010). Resilience thinking: integrating resilience, adaptability and transformability. *Ecology and Society* 15 (4): 9.

31 Ojala, M. et al. (2021). Anxiety, worry, and grief in a time of environmental and climate crisis: a narrative review. In: *Annual Review of Environment and Resources*, vol. 46 (ed. A. Gadgil and T.P. Tomich), 35–58. Palo Alto: Annual Reviews.

32 Pihkala, P. (2019). *Climate Anxiety*. Suomen mielenterveysseura (MIELI Mental Health Finland).

33 Franzke, C.L. (2014). Nonlinear climate change. *Nature Climate Change* 4 (6): 423–424.

34 Kerr, R.A. (2008). Climate tipping points come in from the cold. *Science* 319 (5860): 153.

35 Lenton, T.M. (2011). Early warning of climate tipping points. *Nature Climate Change* 1 (4): 201–209.

36 Lenton, T.M. (2012). Arctic climate tipping points. *Ambio* 41 (1): 10–22.

37 Scheffer, M. (2020). *Critical Transitions in Nature and Society*. Princeton University Press.

13

Risking Life: Basics of Biological Diversity

Consider all this; and then turn to this green, gentle, and most docile earth: consider them both, the sea and the land; and do you not find a strange analogy to something in yourself?

Herman Melville, Moby Dick 1851 [1]

The diversity of life on Earth has waxed and waned since its origins about 3.7 billion years ago, (though earlier dates for its origin have been proposed) [2, 3]. This overall trend of increasing diversification over time is punctuated by periods of decline, although such periods are short compared to the long sweep of the geological ages.

A classic question in biodiversity studies is: Why are there so many species? Rephrasing the question, we can ask: Why is there not only one? Though the world's species diversity is in no danger of being reduced to a single species, nonetheless, the seeming collapse of biodiversity in recent centuries, and accelerating over the past few decades, has provoked many important questions. Among them is this: Will this have catastrophic implications for human welfare?

Variability in beetles; Staatliches Museum für Naturkunde Karlsruhe, Germany. Shared under the Creative Commons Attribution-Share Alike 3.0 Unported license. *Source:* H. Zell/Wikimedia Commons / CC BY-SA 3.0.

Over the past 550 million year or so, there have been at least five periods of mass extinction. We are now in the sixth such period of rapid decline, a claim that is based upon a growing body of credible evidence. Not only are we living through a collapse in diversity but the decline is accelerating, and, furthermore, on this occasion, the human transformation of the globe is the primary driver of a spasm of extinction [4]. What are the implications of this loss for the functioning of the planet, and upon the ecological largesse of which we are completely dependent? Will this loss have *catastrophic* consequences for human life – catastrophic in the sense that we discussed in the previous chapter? On the other hand, is it possible that at least some of the diversity of life is redundant – that is, is not required for functioning – and thus, though the loss of life's diversity may be greatly elevated – and this is certainly an ethical catastrophe – it is a loss that can that be endured from the perspective of human material welfare?

In this and the following chapter, we have an opportunity to review some of the general conceptual frameworks we introduced earlier in the book, especially those related to complex adaptive systems, and their integrity and functioning. Additionally, this chapter lays the conceptual groundwork for examining the causes and consequences of biodiversity loss in the Anthropocene, a matter that we take up in the subsequent chapter.

13.1 Why Is there Not Just One Species of Living Thing?

Why is there not just a single species on Earth: a living entity so wily that it is capable of thriving in all conceivable circumstances? Such a putative super-species – referred to in thought experiments as a "Darwinian demon" – cannot exist because every living being is finite and thus is constrained in its environmental interaction by its physiology, energy acquisition abilities, and genetics. Thus constrained, when an organism maximizes any one trait (e.g. belowground nutrient capture) it incurs costs with respect to other traits (e.g. photosynthetic capture of energy). Because such trade-offs in trait specialization are prevalent in nature, our would-be demon – assuming it conforms to same laws of variance and hereditary as any other creature – *must* eventually be outcompeted in at least some respects by members of another population, and thus passes its fiendish reign as world usurper [5, 6]. The diversity of life from the perspective of this scenario is the inevitable manifestation of the interaction between finite life and environment. Given the nature of the organism, interacting with its environment under well-characterized evolutionary and ecological constraints diversity emerges.

Another way of conceiving patterns in the diversity of life (as being somewhat inevitable) is to consider the following question: If the number of species on earth were reduced to a single one, would species diversity eventually re-emerge? As far as we know, a reduction of the diversity of life to a single species has never occurred. However, on several occasions mass extinction events have indeed reduced diversity to a dramatic extent. After each of the previous five mass extinction events – the most severe of which occurred at the end of the Permian 245 million years ago, an event that eliminated 54% of marine families and perhaps as many as 96% of all marine species – diversity rebounded, in most cases after a lapse of several million years [7, 8]. The continued renewal of the evolutionary diversification of life, at the genetic, population, species level and beyond is one of the most general patterns in the history of life [9, 10].

The total number of species that exists at any one time reflects a balance between the inexorable forces of species production (speciation) and extinction, rates of both of which fluctuate [11]. When extinction rates vastly exceed speciation this is referred to as a mass extinction event, regardless of what forces drive the loss [12]. In this and the next chapter we examine the evidence that the human transformation of the globe is accelerating extinction rates to the extent that a mass extinction event is now occurring. This phenomenon has been called the Holocene extinction (also referred to as the Sixth extinction or Anthropocene extinction) [13, 14].

Before addressing questions related to anthropogenic impacts on the diversity of life, we needed to understand the basic concepts of biodiversity science. This is the primary concern of this chapter.

13.2 Diversity of Life

Life comes in an astonishing array of forms (for an excellent searchable web site with information and images of a wide variety of species see *The Catalogue of Life* project (https://www.catalogueof-life.org) [15, 16]). That all of life descends from a common point of origin is the dominant theory (though there may have been multiple originating events with only one leaving behind descendants) [17, 18]. Living things are thus monophyletic – that is, are derived from this common ancestor – and, subsequent to its origin, life has diversified into the vast range of organisms that have existed over the past 3.7 billion years [19]. All of these organisms form parts of a complex system placed within hierarchically arrayed taxonomic categories [20, 21]. Typically, an organism – a fundamental unit, or part, in this system – is classified as a member of a species, and species in turn are placed into progressively more encompassing taxa: genus (plural genera), family, order, class, phylum (plural phyla; in the case of plants, the term "division" is used rather than phylum), kingdom, and domain. For example, modern humans are classified as *Homo sapiens*, Genus Homo, Family Hominidae, Class Mammalia, Order Primates, Phylum Chordata, Kingdom Animalia, (or Domain Eukaryota, depending upon which higher-level taxonomic scheme is preferred). In addition to these taxonomic categories, there are levels below the species (subspecies, genetically distinct populations) that are important for conservation. For example, contemporary humans are typically assigned the subspecies *sapiens*, making us *Homo sapiens sapiens*. This distinguishes us from archaic modern humans, *Homo sapiens idaltu* [22].

In the above schema, we can recognize that our taxonomy and classification of organisms is a hierarchical system: each of the parts can be assembled into progressively larger units (species as the "parts" in genera; genera in families, and so on). This can be described as a genealogical hierarchy of nature – that is, a sequence of parts connected by reproduction extended through time [23].

There is no one universally accepted schema for aggregating all organisms into the ultimate, most all-encompassing levels of this system. One novel schema for doing so, often utilized these days, assigns all living entities into one of the three general domains: bacteria, archaea, and eukaryotes. This system was proposed by biologist Carl Woese; it is based upon the analysis of ribosomal RNA (and on the core genes involved in protein translation) [24]. (Ribosomes are tiny components in all cells that are associated with the synthesis of protein.) More recent genetic analysis suggests that the eukaryotes arose from a partnership of the archaea and bacteria, and this work supports a radical proposal that all species can be classified in one of two domains (bacteria, archaea) [25]. However, other systematists disregard the category of "domain," preferring to classify all of life into the following groups: two prokaryotic kingdoms Archaea (Archaebacteria) and Bacteria (Eubacteria), as well as the eukaryotic kingdoms Protozoa, Chromista (a relatively newly named kingdom that include algae and many organisms formerly classified as protozoans), Fungi, Plantae, and Animalia [24, 26].

Regardless of which particular classification schema is utilized, all Linnaean taxonomic systems (that is, ones that follow the pioneering work of Carl Linnaeus [1707–1778]) are hierarchical. As one ascends from the level of species to the domains, each new level encompasses, and is more complex than, the one below: there are fewer genera than species, fewer families than genera, and so on, up to the singularity of life itself. The highest taxonomic levels are more conservative than lower ones in the sense that they endure over very long periods. The domains and kingdoms that exist today persisted through all five major mass extinction events (and might be expected to endure the current spasm of extinctions) whereas species, general, families, etc. are more vulnerable, and mass extinctions are often assess at the familial level [27].

Discussions of the classification of life can seem arcane, but even our brief overview reveals patterns that accords well with insights from complex adaptive systems theory. In addition to each level in the genealogical hierarchy of life integrating the levels below, each level will have its own set of defining characteristics. This is why, for example, there is such interest in the study of now extinct members of the genus to which we below (*Homo*): to study the totality of these species can disclose information on what it is to be human. The existence of entities above the species level (genus, family, and so on) depends upon the existence of the species level. As we proceed in the sections below to a discussion of biodiversity – that is, the consideration of the diversity of life in all its manifestations – we will see that both the *ecological hierarchies* (that is, local systems nested into regional and global systems) and *functional hierarchies* (species in ecosystems) in which living things participation, can be mutually influential. Species influence ecosystems, but are in turn constrained by them.

Recognizing the complexity of the genealogical hierarchy of life (species, genus, family etc.), a critical and pragmatic first step in contemporary evaluations of the state of biodiversity loss is selecting the appropriate taxonomic level for assessment and conservation action. Considerable attention is given in conservation policy debates to the species level [28, 29] and public concern for nature protection often revolves around species, in particular the protection of rare species (e.g. Javan rhinos, Mountain gorillas, Tapanuli orangutans) [30, 31]. Therefore, it is worth discussing in some detail what is meant by a species. Not only is there an ongoing debate about what, exactly, a species is– this is called the "species problem" – there is concern among some conservation biologists that focusing on species is to the detriment of global conservation efforts.

13.3 What Is a Species?

All languages it seems place organisms into hierarchically arranged taxa [32]. Furthermore, the perception of many taxonomic categories appear to be robust across cultures: from an all-inclusive category of an ancestral living being (in folk taxonomy research this is often called the "unique beginner"), down to local varietals at taxonomic levels that exist below the level of species. Of these shared groups, the genus and species seems to be the most commonly used in informal systems; genera are often designated by a single word (technically, a single lexeme), like "oak" or "robin," and species which are labeled with two words (a binomial in Linnaean taxonomy): like bur oak (*Quercus macrocarpa*), and European robin (*Erithacus rubecula*) [32]. The fact that the word "human" is used to refer to a single species is simply because we are the last remaining of a formerly more diverse genus [33]. Since aggregating organisms into categories that supposedly represent close affinities is a near universal cultural activity, folk taxonomy, especially in the identification of genus and species, can be helpful in scientific work and vice versa. That is, scientific conservation at the species level may serve local resource management needs. For example, fishers in the Amazon basin discriminate species into groupings similar to those of Linnaean ichthyologists; for this reason, local experts are often recruited to work as parataxonomists in formal research endeavors [34, 35].

The concordance of species in folk and Linnaean taxonomies, however, is not always precise: often a species name that is used in vernacular settings corresponds to a genus name, or even a subspecies in biological taxonomies [32, 36]. Despite the absence of a full concordance between vernacular and the Linnaean means of naming organisms, it seems as if the "species" category in some form is almost universally employed for the purposes of both practical engagement with nature (for example, fishing or farming) and in scientific work. This focus upon readily apparent discontinuities in the appearance and the behavior of organisms generally validates the opinion that the species is a both nonarbitrary and practical unit. Because of the cross-cultural preoccupation with naming creatures, the species is often seen as "real" rather than as "nominal" [34]. By species-realism is meant a view that species exist as distinct entities in nature independently of the minds or acts of scientists. They are not merely a subjectively chosen and convenient ways of naming things – a perspective called species nominalism. Nominalism points to a continuum in the natural world, and regards the distinction between species as merely a convenience.

Even Darwin has been recruited into this debate over the so-called "species problem," and is seen as being a defender of one view or the other depending upon which author is consulted. Darwin certainly seems to have conceded the arbitrariness of the species concept as can be seen in the following:

> No one definition has as yet satisfied all naturalists; yet every naturalist knows vaguely what he means when he speaks of a species.

Elsewhere Darwin wrote:

> From these remarks it will be seen that I look at the term species, as one arbitrarily given for the sake of convenience to a set of individuals closely resembling each other, and that it does not essentially differ from the term variety, which is given to less distinct and more fluctuating forms [18]

Despite Darwin's seeming ambivalence on this question, some historians of science (e.g. David N. Stamos) have stoutly defended him against a contention that he was a species nominalist (that is, the idea that species are a convenience and not real) [37]. As Stamos summarizes: "there appears no good reason. . .to believe that he would have denied the reality of the species category, although he may not have had a sufficiently clear idea in order to define it" [37]. Whether or not we take a firm position on this difficult, and seemingly unresolved, philosophical problem, the following two things seem undeniable: no single definition of species will satisfy all scientists, and the basis for divergent definitions are rooted in nontrivial distinctions.

13.4 A Diversity of Species Concepts

Since the category of species, as a fundamental unit of the complex system of life, serves a role in the economic, recreational, and scientific life of humans, it may not be surprising that there is no single concept of species that fully satisfies all uses. Having recognized early in the book that many core ideas in environmental disciplines are vague, and yet need to be sharply defined for doing scientific work, the fact that multiple definitions of species exist will come as no surprise. Indeed, formulating definitions may be the product of research, at least in the early stages of a systematic study.

The species concept is a fundamental unit in systematics, evolutionary studies, and conservation science; these different branches of science have formulated more than 20 species concepts employed for this variety of purposes [38, 39]. The ones, however, that are in most common use revolve around the following distinctions.

Firstly – and most practically – a species can be defined based upon variation in its anatomical, physiological, and behavioral characters. A character in taxonomy refers to those attributes of organisms that seem fixed (or unchanging) in that taxon [40]. Identifying these characteristics of organisms – especially of morphological characters – provides a very practical approach for taxonomists identifying species in the field or based upon taxonomic collections (for example, in museums). Examples of fixed characters used in taxonomy will include diagnostic color patterns in birds, size and body shape in many organisms, and even wing venation in most insects, since these patterns are reasonably constant. Morphologically based species concepts can be implemented by means of a close examination of the characters of those individual organisms chosen to serve as exemplars of species. With information gleaned from a close inspection of specimens, a taxonomist can apply a numerical analysis to clusters of these characters to discriminate one species from another, and to evaluate possible relationships. Such morphological-based taxonomy works especially well when the entities being examined are discrete and characters are clearly disjunct from those of other populations. They become harder to apply when boundaries between the characters of different populations, or even closely related species, are vague. Species concepts associated with morphology are sometimes referred to as the *phenetic* ones, that is, they are based upon overall similarities of characteristics (phenetic derives from the Greek for appearance) [41, 42]. The use of these techniques and their implementation does not rely upon any knowledge of the function of the observed characters, or the evolutionary relationship between species [43].

The form and organization of structures of an organism is referred to as its **morphology**. The process of natural selection operates on variations of morphology – these are the "characters" studied by taxonomists.

A second cluster of species concepts is based explicitly upon attempts to discern distinct lineages of organisms based upon the biological process of reproduction (which produce lines of descent with modification). Viewed in this way, species are a product of evolutionary change; that is, a species is an identifiably distinct evolutionary unit. The best known among species concepts such as these is the biological species concept (BSC) [44]. The BSC is a multidimensional concept because it integrates the study of variation in morphology and behavior, along with a consideration of both the geographical distribution over large areas and the genetic linkages between populations. The genetic linkages between individual within species are reinforced by the "reproductive isolation" of each species – its inability to reproduce and produce viable offspring. The systematist and renowned biologist Ernst Mayr (1904–2005), defines the BSC as follows: "I define biological species as groups of interbreeding natural populations that are reproductively isolated from other such groups." However, the concept has been heavily criticized on both practical grounds (who will test, for most species, whether individuals of opposite sex can potentially interbreed to produce fertile young?) and as being unnecessary for evolutionary theorizing [41]. The concept applies more readily to sexually reproducing organisms, though many species (especially prokaryotes) do not reproduce in this way.

Other systematists have abandoned the multidimensional BSC for ones based exclusively upon attempts at circumscribing the evolutionary lineage of a group of organisms. Such species concepts include the evolutionary species and the phylogenetic species concepts [45]. Focusing on evolutionary lineages allows for a discrimination between distinct clusters of organisms: a species, in the phylogenetic sense, may be viewed as the largest aggregate of individual organisms that evolve as a unit [46].

As should be apparent from the above paragraphs, the literature on the "species problem," – that is, the attempts to make something concrete out of a vague concept – is immense and its production shows no signs of relenting any time soon [47–49]. Though these attempts to understand what exactly a species is, seem, perhaps, of more immediate relevance to the fields of taxonomy and evolutionary studies, it is also a crucial matter for the conservation of life on earth. This is because the species remains the privileged unit of conservation biology [50]. The meaning of, and the practical identification and classification of species – especially ones that rare and vulnerable – can affect investment in research upon endangered taxa, and the legal protection afforded species [51]. Conservation biologists Stephen Garnett and Les Christidis express the concern that the vagueness of species concepts undermines the effectiveness of efforts to halt biodiversity loss on the global scale. They urge that, at the very least, a governing body should oversee the imposition of a uniform taxonomic code to guide conservation efforts [51]. Ultimately, we may need to accept that no one species concept will fit all the needs to which it is put. Just as taxonomic and evolutionary studies have done, conservation biologists and managers must accept that though species may form a fundamental unit in the discipline, as a category, each species will always be an expression of dynamic demographic, geographical, ecological, and evolutionary processes [52, 53].

Since the species remains, and is likely to continue on as a fundamental unit in conservation biology and in efforts to protect or restore the diversity of life on Earth it seems important, therefore, to know, just how many such units there are. As we shall see, this has turned out to be an extremely difficult number to find out.

13.5 How Many Species?

Translated into systems terms, estimates of the number of species of earth is an attempt to account for the number of relevant parts, the interactions of which produce the behavior of the global living whole. Estimating the global tally of species is an endeavor of some antiquity and remains a matter of considerable ongoing debate [54]. For example, writing in the seventeenth century, the English naturalist John Ray (1627–1705) opined ". . .in consequence of having discovered a greater number of English moths and butterflies I am induced to consider that the total number of British insects might be about 2000; and those of the whole Earth 20 000"; the great Linnaeus suggested that about 4400 species of animals existed globally [54], though by the writing of the 1758 edition of *Systema Naturae* he included 9000 species [36].

In the centuries following Linnaeus, as ever more species were discovered, there was an inevitable acceptance that global richness was much greater that previously recognized and that the total was surely in the millions. The estimates of both Ray and Linnaeus were clearly considerably below the mark. As a result of more than two and a half centuries of taxonomic effort there are now as many as 1.9 million species described by scientists and assigned binomial names [55, 56]. (Other tallies of named species are lower: perhaps only 1.5 million species have been described).

The variability of these estimates reflects the enormity of the task of keeping track of the total species count [57]. How long it will ultimately take to name all the species on the planet depends, of course, both upon the rate at which the task of discovering and describing new species can proceed, and upon how many undescribed species remain. The task of describing species can be a very slow one: the time lag between specimen acquisition and the publication of a scientific description can often be more than two decades [58]. Along with the inherently painstaking nature of the job, there have been widespread expressions of concern about the underfunding of taxonomy and the suspicion that the basic sciences of taxonomy and systematics are being supplanted by newer areas of biology, such as genetics and cell biology (both of which may contribute to taxonomic work) [59]. Despite such expressions of concern about declining funding and expertise, recent analysis suggest that there are more taxonomists describing species than ever – as many as 40 000 taxonomists are working worldwide – and their ranks are increasing faster than the current rates of extinction [60]. Though evidence of an increase in taxonomic expertise is heartening, whether the number is adequate for the task of describing all species before many are gone forever remains to be seen.

The fact is this: we cannot be confident that the number of species on Earth is known to the *nearest order of magnitude* – we might be off by millions or tens of millions [61, 62]. The lack of information we have about this matter is noted with a tone of incredulity by scientists reviewing the question. For example, ecologist R.M May (1936–2020) writes:

> It is a remarkable testament to humanity's narcissism that we know the number of books in the US Library of Congress on 1 February 2011 was 22,194,656, but cannot tell you – to within an order-of-magnitude – how many distinct species of plants and animals we share our world with [63].

It remains the case that answering the question about global species richness is complicated by the unlikeliness of discovering a unified species concept that puts estimates of diversity across taxa on an equal footing [49]. As May and many others note, the incompleteness occurs partly because we now know that much of the "missing diversity" is comprised of species from relatively understudied taxa of small body size, and living in inaccessible habitats [57, 59, 62, 64–66].

Though uncertainties about global species diversity have long been recognized, until the early 1980s, however, there remained some optimism that the task was tractable and perhaps even nearing completion (there were, according to estimates in that decade, perhaps, two or at most three million species [67]). However, in 1982 in a note published in *The Coleopterist Bulletin* – a specialized journal that only rarely attracts global attention – the American taxonomist Terry Erwin revolutionized estimates of diversity. Erwin conducted an insightful study in a Panamanian forest [68]. In his brief paper – it is only two pages long – Erwin presented the results of his releasing an insecticidal fog into the canopy of a single tropical tree species *Luhea semannia*. Thousands of specimens drizzled from the treetops into Erwin's collectors far below. From these samples, Erwin identified 163 beetle species that he argued were specialist feeders on that one species of tropical tree. From this modest sample, Erwin then proceeded to extrapolate, in a relatively straightforward way, to estimates of global diversity. Assuming that all tropical tree species had equivalent numbers of specialists, and since there are 50 000 tropical tree species in all, that would yield a conservative estimate of 8 150 000 tropical beetles. Since beetles represent approximately 40% of all arthropods, this would mean that there are at least 20 million tropical arthropods. Adding species from all other nonbeetle taxa, and those from nontropical realms, would add extra millions to the tally of species, giving a grand total of over 20 million species: perhaps there are as many as 30 million. In this single paper, Erwin had transformed the task of accounting for the diversity of life from a difficult one, to one that was next to impossible at the rate at which taxonomists describe species.

A paper by American entomologist, Terry Erwin, published in 1982 revolutionized global biodiversity studies. That paper investigated the previously unknown hyperdiversity of beetle species in the crowns of tropical forests. This work stimulated renewed interest in previously understudies "biotic frontiers" like tree crowns, soil, and the deep ocean. *Source:* Mikenorton/Wikimedia commons.

Erwin's estimate of global diversity undeniably stimulated a great interest in the global species tally although the details of his method – extrapolating from small but well characterized samples in geographically restricted areas up to the global scale – has come in for criticism. His estimate is based, as we have seen, upon an intensive localized sampling of trophic specialists, but has been regarded by his critics as extreme [69]. Criticism has concentrated, in particular, on the sensitivity of his calculations to assumptions made about the host-specificity of beetles on *Luhea semannia*, though Erwin has subsequently defended his conjecture [69, 70].

Since Erwin's calculation in his 1982 paper were made, his method and variants upon it have been employed by other taxonomists [36]. Additionally, a range of other estimation techniques have been employed: these include estimating based upon (1) calculating the ratios of extremely well-known taxa (for example, butterflies in Great Britain) compared to unknown arthropod groups, (ii) estimating discovery rates for new species, (iii) evaluating the educated guesses of taxonomic experts, and (iv) using predictions based on body size (generally the small the taxa, the greater the diversity) [71]. None of these techniques is definitive in and of themselves, but the results from applying several of them have converged, and allow for confidence limits to be placed around estimates. We may now conjecture that there are 5.5 million species of insects (the range is 2.6–7.8 million), and 6.8 million species of terrestrial arthropods (range 5.9–7.8 million). As a result, global species tallies of eight million animals and 350 000 plants now seem quite plausible [36, 68, 72–75]. According to biologist Camilo Mora and his colleagues, for example, "86% of existing species on Earth and 91% of species in the ocean still await description" [60]. Although these estimates are certainly higher than was supposed to be the case in the 1980, they are not drastically so – the task may be achievable after all, though it would still take a monumental investment in taxonomic infrastructure [59].

Despite the new information gleaned by asking questions about the global tally of life, knowing precisely what the species count is, may not, in itself, be all that important. This point is made emphatically by Terry Erwin, who writes: "Finally. . . perhaps it does not really make any difference exactly how many species there are. The world is losing them at an astounding rate along with their habitats, and we humans are responsible" [70]. Echoing this sentiment is the reminder by ecologist Kevin Gaston – a critic of Erwin's approach to estimating global diversity – who wrote, "Whether there are five, ten, fifteen, or more million species of insects in no way alters their individual significance, nor the significance of the crisis with which they are currently faced" [76].

13.6 Biodiversity and Its Measurement

It is becoming quite clear in global conservation biology discussions that, depending upon the objectives of a given conservation intervention, targeting species may be appropriate. Nonetheless, there are many conservation objectives, the achievement of which will need to focus on different levels of the taxonomic hierarchy as well as upon the ecological and functional hierarchies in which they participate. For example, when the protection of functional aspects of diversity is regarded as the priority, or the protection of entire landscapes seems appropriate, the species level may be less relevant. Functional diversity, for instance, oftentimes does not readily map onto neat taxonomic categories. For example, nitrogen fixation in soil is performed by diazotrophs, that is, microbes including bacterial and archaeal species that can fix atmospheric nitrogen gas [77–79]. The conservation of an ecological function can often be achieved without explicitly protecting individual species.

Recent appeals for a more eclectic approach to selecting conservation targets in the Anthropocene – both below and above the species – are based upon the recognition that important variation occurs at levels above as well as below the species. Since 2008, the International Union for Conservation of Nature (IUCN), a flagship conservation group, initiated the development of a Red List of threatened and endangered ecosystems [80]. This complements the IUCNs efforts to identify vulnerable species.

A consideration of the diversity of life from a genetic and taxonomic point of view is, as we have seen, crucial to understanding the history of life, and the patterns of speciation and extinction. Recent efforts to both conserve life and protect human welfare rely to an increasing extent upon the development of powerful new tools for genetic analysis used to refine taxonomic analysis. However, the study of the diversity of life and the tools required becomes more complex still when the variety of communities, ecosystems, the spatial distribution of diversity and functional diversity are taken into account.

The composition, structure, and function of life are found at multiple levels of organization. In a typical forest, not only are there multiple species but variation exists at the genetic level (between individuals and species) at the level of plant (and animal) functional and structural group. Looking at this depiction of a forest can you enumerate the sources of variation at the compositional, structural and functional level in this system? How would you measure each type of diversity? *Source:* Elke Freese. See: https://commons.wikimedia.org/wiki/File:Stockwerke_wald.png.

Biodiversity refers to the variety of life and its processes [81]. The term encompasses, as we have intimated above, all of the sources of variations in life. It explicitly includes both evolutionary and ecological processes: indeed, biodiversity can be considered the manifestation (assessed in terms of diversity) of all evolutionary and ecological patterns. A useful scheme to classify the components of biodiversity and derive measurable indicators of biodiversity is presented by the conservation biologist Reed Noss [82]. He identifies the following important metrics by which biodiversity can be assessed.

1) The variety of life's *composition* (an enumeration of genetic, ecological, and spatial components of diversity).
2) The variety of life's *structure* (the extent to which populations, habitats, or landscapes are heterogeneous in their physical organization within local habitats and the larger landscape).
3) The variety of life's *function* (the manner in which the genetic, demographic, interspecies, landscape processes operate in executing their roles) of the concept of biodiversity.

These three attributes of biodiversity (compositional/genealogy, structural/ecological, and the functional/process attributes) can be organized into a hierarchical scale. The scale proposed by Noss is fourfold; thus, the compositional, structural, and functional aspects of biodiversity can be measured at the scales of genetic, population-species, the community-ecosystem, and the regional-landscape scale. Noss's matrix includes twelve measureable attributes in all (3 attributes × 4 points in a hierarchical scale) by which biodiversity can be assessed. Indicator variables and monitoring tools for each of the twelve elements have been developed (for example, patterns of species distribution can be evaluated at the landscape scale; at the genetic scale, functional diversity can be measured by assessing genetic drift, gene flow, mutation frequency, and selection intensity).

This complexity underscores that challenges involved in undertaking a full biotic inventory; it may be unlikely that all will be include in any one scheme. However, as in all such hierarchies, the lower levels – local genetic and population dynamics – can be subject to very rapid change and often operate on small spatial scales, whereas the upper levels – ecosystem and landscape – are slower and more stable. The significance of this is that depending upon the scope of a given conservation project – the time scale of its implementation and the spatial extent of its responsibilities – not all aspects of biodiversity may need to be assessed. On the other hand, unless all components of biodiversity at all scales of assessment are, to some extent at least, incorporated into conservation planning, then long-term success in protecting, conserving, and restoring the diversity of life on Earth may be in jeopardy.

It is this fairly complex notion of biodiversity that will form the basis for our deliberations in the next chapter about the role of human-wrought changes in unraveling the fabric of life – a phenomenon referred to as the Sixth Extinction.

13.7 Coda

Ignorance about the diversity of life on Earth exists despite the fact that interest in animals may be among humanity's most perennial concerns, as can be gleaned from, for example, Indonesian rock art dating back to almost forty thousand years ago, and from classificatory schemes (of plants and animals) of very great antiquity [83, 84]. The diversity of life may be an enduring preoccupation, yet questions concerning its fate, and the implications of its loss on the functioning of earth systems, and on human welfare are not easily answered. Despite the extreme uncertainties about the composition, structure, and function of biodiversity at various scales, we, nonetheless, can be confident of a number of things: contemporary extinction rates are elevated, biotic communities are being homogenized, and ecosystem function is being impaired in consequential ways [85–87]. Whether these diminishments will lead to global catastrophe is uncertain, but without a doubt, this still relatively underexplored issue, will have grave consequences for human flourishing.

References

1 Parker, H. and Hayford, H. (2001). *Moby-Dick (Norton Critical Editions)*. New York: WW Norton.

2 Doolittle, W.F. (1999). Phylogenetic classification and the universal tree. *Science* 284 (5423): 2124–2128.

3 Javaux, E.J. (2019). Challenges in evidencing the earliest traces of life. *Nature* 572 (7770): 451–460.

4 Kolbert, E. (2014). *The Sixth Extinction: An Unnatural History*. A&C Black.

5 Kneitel, J.M. and Chase, J.M. (2004). Trade-offs in community ecology: linking spatial scales and species coexistence. *Ecology Letters* 7 (1): 69–80.

6 Law, R. (1979). Optimal life histories under age-specific predation. *The American Naturalist* 114 (3): 399–417.

7 Erwin, D.H. (1990). The end-Permian mass extinction. *Annual Review of Ecology and Systematics* 21 (1): 69–91.

8 Twitchett, R.J. (1999). Palaeoenvironments and faunal recovery after the end-Permian mass extinction. *Palaeogeography, Palaeoclimatology, Palaeoecology* 154 (1, 2): 27–37.

9 Wilson, E.O. (1999). *The Diversity of Life*. WW Norton & Company.

10 Attenborough, D. (1979). *Life on Earth: A Natural History*. Boston: Little, Brown.

11 Ricklefs, R.E. (2014). Reconciling diversification: random pulse models of speciation and extinction. *The American Naturalist* 184 (2): 268–276.

12 Benton, M.J. (1995). Diversification and extinction in the history of life. *Science* 268 (5207): 52–58.

13 Dasgupta, P. and Ehrlich, P. (2019). Why we're in the sixth extinction and what it means to humanity. In: *Biological Extinction: New Perspectives, Forthcoming* (ed. P. Dasgupta, P.H. Raven and A. McIvor), 262–284. Cambridge, UK: Cambridge University Press.

14 Turvey, S.T. and Crees, J.J. (2019). Extinction in the Anthropocene. *Current Biology* 29 (19): R982–R986.

15 Maddison, D.R., Schulz, K.-S., and Maddison, W.P. (2007). The tree of life web project. *Zootaxa* 1668 (1): 19–40.

16 Cachuela-Palacio, M. (2006). Towards an index of all known species: the catalogue of life, its rationale, design and use. *Integrative Zoology* 1 (1): 18–21.

17 Raup, D.M. and Valentine, J.W. (1983). Multiple origins of life. *Proceedings of the National Academy of Sciences* 80 (10): 2981–2984.

18 Darwin, C. (1996). *The Origin of Species: Oxford World's Classics*. Oxford University Press.

19 Pearce, B.K.D. et al. (2018). Constraining the time interval for the origin of life on earth. *Astrobiology* 18 (3): 343–364.

20 McNeill, J. et al. (2012). *International Code of Nomenclature for algae, fungi and plants (Melbourne Code)*, vol. 154. Königstein: Koeltz Scientific Books.

21 Ride, W. (1999). *International Code of Zoological Nomenclature*. International Trust for Zoological Nomenclature.

22 White, T.D. et al. (2003). Pleistocene homo sapiens from middle awash, Ethiopia. *Nature* 423 (6941): 742–747.

23 Salthe, S.N. (1985). *Evolving Hierarchical Systems*. Columbia University Press.

24 Woese, C.R., Kandler, O., and Wheelis, M.L. (1990). Towards a natural system of organisms: proposal for the domains archaea, bacteria, and eucarya. *Proceedings of the National Academy of Sciences* 87 (12): 4576–4579.

25 Williams, T.A. et al. (2013). An archaeal origin of eukaryotes supports only two primary domains of life. *Nature* 504 (7479): 231–236.

26 Ruggiero, M.A. et al. (2015). A higher level classification of all living organisms. *PLoS One* 10 (4): e0119248.

27 Purvis, A. (2008). Phylogenetic approaches to the study of extinction. *Annual Review of Ecology, Evolution, and Systematics* 39: 301–319.

28 Arponen, A. (2012). Prioritizing species for conservation planning. *Biodiversity and Conservation* 21 (4): 875–893.

29 Pickett, S.T., Parker, V.T., and Fiedler, P.L. (1992). The new paradigm in ecology: implications for conservation biology above the species level. In: *Conservation Biology* (ed. P.L. Fiedler and S.K. Jain), 65–88. Springer.

30 Czech, B. and Krausman, P.R. (1999). Research notes public opinion on endangered species conservation and policy. *Society & Natural Resources* 12 (5): 469–479.

31 Angulo, E. et al. (2009). Fatal attraction: rare species in the spotlight. *Proceedings of the Royal Society B: Biological Sciences* 276 (1660): 1331–1337.

32 Berlin, B., Breedlove, D.E., and Raven, P.H. (1973). General principles of classification and nomenclature in folk biology. *American Anthropologist* 75 (1): 214–242.

33 Tattersall, I. (2012). *Masters of the Planet: The Search for our Human Origins*. St. Martin's Press.

34 Coyne, J.A. and Orr, H.A. (2004). *Speciation*, vol. 37. Sunderland, MA: Sinauer Associates.

35 Begossi, A. et al. (2008). Are biological species and higher-ranking categories real? Fish folk taxonomy on Brazil's Atlantic forest coast and in the Amazon. *Current Anthropology* 49 (2): 291–306.

36 Eichhorn, M. (2016). *Natural Systems: the Organisation of Life.* Wiley.

37 Stamos, D.N. (1996). Was Darwin really a species nominalist? *Journal of the History of Biology* 29 (1): 127–144.

38 Mallet, J. (2001). Species, concepts of. *Encyclopedia of Biodiversity* 5: 427–440.

39 Hey, J. (2001). The mind of the species problem. *Trends in Ecology & Evolution* 16 (7): 326–329.

40 Davis, J.I. and Nixon, K.C. (1992). Populations, genetic variation, and the delimitation of phylogenetic species. *Systematic Biology* 41 (4): 421–435.

41 Sokal, R.R. and Crovello, T.J. (1970). The biological species concept: a critical evaluation. *The American Naturalist* 104 (936): 127–153.

42 Desalle, R. (2006). What's in a character? *Journal of Biomedical Informatics* 39 (1): 6–17.

43 Ridley, M. (1986). *Evolution and Classification: The Reformation of Cladism.* London, New York: Longman.

44 Mayr, E. (2000). *The Biological Species Concept. Species Concepts and Phylogenetic Theory: A Debate,* 17–29. New York: Columbia University Press.

45 Cracraft, J. (1987). Species concepts and the ontology of evolution. *Biology and Philosophy* 2 (3): 329–346.

46 Wiley, E.O. (1978). The evolutionary species concept reconsidered. *Systematic Zoology* 27 (1): 17–26.

47 Pavlinov, I.Y. (2021). *Conceptualization of the species problem, in Species in Biology: Theory and Practice,* 69–85. Kyiv: Ukrainian Theriological Society & NMNH NAS of Ukraine.

48 Thiele, K.R. et al. (2021). Towards a global list of accepted species I. Why taxonomists sometimes disagree, and why this matters. *Organisms Diversity & Evolution* 21: 615–622.

49 Zachos, F.E. (2016). *Species Concepts in Biology,* vol. 801. Springer.

50 Rojas, M. (1992). The species problem and conservation: what are we protecting? *Conservation Biology* 6 (2): 170–178.

51 Garnett, S.T. and Christidis, L. (2017). Taxonomy anarchy hampers conservation. *Nature* 546 (7656): 25–27.

52 Thomson, S.A. et al. (2018). Taxonomy based on science is necessary for global conservation. *PLoS Biology* 16 (3): e2005075.

53 Raposo, M.A. et al. (2017). What really hampers taxonomy and conservation? A riposte to Garnett and Christidis (2017). *Zootaxa* 4317 (1): 179.

54 Ødegaard, F. (2000). How many species of arthropods? Erwin's estimate revised. *Biological Journal of the Linnean Society* 71 (4): 583–597.

55 Costello, M.J., Wilson, S., and Houlding, B. (2012). Predicting total global species richness using rates of species description and estimates of taxonomic effort. *Systematic Biology* 61 (5): 871.

56 Chapman, A.D. (2009). *Numbers of Living Species in Australia and the World.* Australian Department of the Environment, Water, Heritage and the Arts.

57 Stork, N.E. (1993). How many species are there? *Biodiversity & Conservation* 2 (3): 215–232.

58 Fontaine, B., Perrard, A., and Bouchet, P. (2012). 21 years of shelf life between discovery and description of new species. *Current Biology* 22 (22): R943–R944.

59 Blackmore, S. (1996). Knowing the Earth's biodiversity: challenges for the infrastructure of systematic biology. *Science (American Association for the Advancement of Science)* 274 (5284): 63–64.

60 Costello, M.J., May, R.M., and Stork, N.E. (2013). Can we name Earth's species before they go extinct? *Science* 339 (6118): 413–416.

61 May, R.M. (1988). How many species are there on earth? *Science* 241 (4872): 1441–1449.

62 May, R.M. (1990). *How many species? Philosophical Transactions of the Royal Society of London Series B: Biological Sciences* 330 (1257): 293–304.

63 May, R.M. (2011). Why worry about how many species and their loss? *PLoS Biology* 9 (8): e1001130.

64 Mora, C. et al. (2011). How many species are there on earth and in the ocean? *PLoS Biology* 9 (8): e1001127.

65 Drew, L.W. (2011). Are we losing the science of taxonomy? As need grows, numbers and training are failing to keep up. *BioScience* 61 (12): 942–946.

66 Scheffers, B.R. et al. (2012). What we know and don't know about Earth's missing biodiversity. *Trends in Ecology & Evolution* 27 (9): 501–510.

67 Stork, N.E. (1997). Measuring global biodiversity and its decline. In: *Biodiversity II: Understanding and Protecting our Biological Resources*, vol. 41, 41–68.

68 Erwin, T.L. (1982). Tropical forests: their richness in Coleoptera and other arthropod species. *The Coleopterists Bulletin* 36 (1): 74–75.

69 Stork, N.E. (1988). Insect diversity: facts, fiction and speculation. *Biological Journal of the Linnean Society* 35 (4): 321–337.

70 Erwin, T.L. (1991). How many species are there?: revisited. *Conservation Biology* 5 (3): 330–333.

71 Stork, N.E. et al. (2015). New approaches narrow global species estimates for beetles, insects, and terrestrial arthropods. *Proceedings of the National Academy of Sciences* 112 (24): 7519–7523.

72 Blackwell, M. (2011). The fungi: 1, 2, 3. . . 5.1 million species? *American Journal of Botany* 98 (3): 426–438.

73 Sweetlove, L. (2011). Number of species on earth tagged at 8.7 million. *Nature*. https://doi.org/10.1038/news.2011.498.

74 Pimm, S.L. et al. (2014). The biodiversity of species and their rates of extinction, distribution, and protection. *Science* 344 (6187): 1246752.

75 Joppa, L.N., Roberts, D.L., and Pimm, S.L. (2011). How many species of flowering plants are there? *Proceedings of the Royal Society B: Biological Sciences* 278 (1705): 554–559.

76 Gaston, K.J. (1991). The magnitude of global insect species richness. *Conservation Biology* 5 (3): 283–296.

77 Devictor, V. et al. (2010). Spatial mismatch and congruence between taxonomic, phylogenetic and functional diversity: the need for integrative conservation strategies in a changing world. *Ecology Letters* 13 (8): 1030–1040.

78 Cadotte, M.W., Carscadden, K., and Mirotchnick, N. (2011). Beyond species: functional diversity and the maintenance of ecological processes and services. *Journal of Applied Ecology* 48 (5): 1079–1087.

79 Sprent, J.I. (1987). *The Ecology of the Nitrogen Cycle*. Cambridge University Press.

80 Keith, D.A. et al. (2015). The IUCN red list of ecosystems: motivations, challenges, and applications. *Conservation Letters* 8 (3): 214–226.

81 Wilson, E.O. (ed.) (1988). *Biodiversity*, 538. Washington, DC: The National Academies Press.

82 Noss, R.F. (1990). Indicators for monitoring biodiversity: a hierarchical approach. *Conservation Biology* 4 (4): 355–364.

83 Aubert, M. et al. (2014). Pleistocene cave art from Sulawesi, Indonesia. *Nature* 514 (7521): 223–227.

84 Lvi-Strauss, C. (1966). *The Savage Mind*. University of Chicago Press.

85 Brook, B., Sodhi, N., and Bradshaw, C. (2008). Synergies among extinction drivers under global change. *Trends in Ecology & Evolution* 23 (8): 453–460.

86 Humphreys, A.M. et al. (2019). Global dataset shows geography and life form predict modern plant extinction and rediscovery. *Nature Ecology & Evolution* 3 (7): 1043–1047.

87 Geisen, S., Wall, D.H., and Van Der Putten, W.H. (2019). Challenges and opportunities for soil biodiversity in the Anthropocene. *Current Biology* 29 (19): R1036-nnnnn1044.

14

Is the Anthropocene Extinction a Global Catastrophe?

We call it Spring though things are dying.

Bill Callahan, "Spring" (2013) [1]

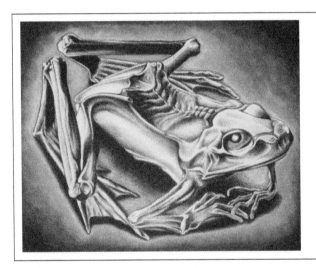

According to estimates published by the IUCN 41% of the over 6000 amphibian species (frogs, toads, salamanders, newts, and caecilians) are threatened with extinction. Most of these species are tropical and occur across small geographic ranges. A combination of ongoing habitat destruction and climate change along with the spread of the virulent disease, chytridiomycosis, makes this ecologically significant vertebrate group uniquely vulnerable to catastrophic collapse. *Source:* M.C. Escher / Wikiart /Public Domain.

Although the exact tally is uncertain, the emerging consensus is that there are many million species of living things on Earth. Taken together, these species represent the totality of life both from a genealogical point of view (that is, the viewpoint that describes ancestral relationships and arranges diversity into the tree of life), as well as from the ecological perspective (in which the hierarchy of life is examined from the scale of individual organisms existing in local systems to the level of life collectively functioning as part of Gaia itself). Species are, to use the language of systems theory, parts of the functioning whole of Earth.

What happens when parts of any system (in this case species, or entire ecosystems) are eliminated from the global whole? As we have discussed earlier in the book, the stability of complex systems – especially at higher levels within a hierarchally organized system – is dependent upon their ability to absorb disturbances that result in the rearrangement or loss of some parts. In the

A Primer on Human Impacts on the Environment: The Conceptual Approach, First Edition. Liam Heneghan.
© 2023 John Wiley & Sons Ltd. Published 2023 by John Wiley & Sons Ltd.

face of disturbance, the biosphere, which is the system that we are examining in this chapter, must be able to continue to perform its functions, while retaining essential structures and system feedbacks. This property is referred to as ecological resilience. Resilience is more readily maintained in systems with strong regulatory feedbacks that operate at multiple scales. High modularity in the construction of ecosystems also contributes to resilience. Modularity is when system parts are organized into semi-autonomous subcomponents – it is useful to remember here the parable of the two watchmakers (see Chapter 5). For any system to be resilient, it will often possess high levels of functional diversity and redundancy. Because of this, resilient and durable systems such as the biosphere can continue to function, by definition, when one part is removed; other components can take over in place of the missing part. Just as we might expect our bodies will endure despite regularly losing cells, so we might expect the stability of Earth systems to endure the loss of species, or even the disappearance or radical alteration of certain habitats.

What happens in circumstance when many parts disappear simultaneously? Under such circumstances, many systems can lose their resilience, fail, and transition to a new state – a state that can be less favorable from a human perspective at least.

In this chapter, we will address several questions related to biodiversity loss. We start by introducing the term extinction: showing that the "naturalness" of extinction has long been known, and forms a conceptual cornerstone of Darwinian evolution. Different forms of extinction are identified and the main drivers of extinction are enumerated. We then contrast the slow background rates of extinction with those that occurred during rare periods in geological time, termed mass extinction events, where extinction rates were elevated. There have been at least five such periods of mass extinction in the past 540 million years. We turn, then, to a discussion of the technical difficulties involved in confirming the extinction of individual species. Despite these difficulties, claims that the scale of contemporary species loss is up to 1000 times faster than the background rate are broadly accepted by the scientific community. We examine the basis of these claims.

What are the consequences of the putative ongoing mass extinction (the so-called Sixth Extinction) for human affairs [2]? Will the continuation of this pattern of loss amount to a catastrophe? In discussing the risks associated with biodiversity decline, we recognize that this loss might be both measurable and severe, though it may not necessarily rise to the threshold of the truly calamitous. In order to distinguish the measurable and severe from the truly catastrophic, we will approach this assessment using similar evaluative tools to the ones we developed in a previous chapter on determining catastrophic risk.

14.1 The Naturalness of Extinction: From George Cuvier to Charles Darwin?

The naturalness of extinction is a hard won scientific fact. The idea that species could disappear from the roster of living beings engaged the minds of the great naturalists of the eighteenth and nineteenth century. Although our contemporary understanding of extinction comes from George Cuvier (1769–1832), the integration of the concept by Charles Darwin in his *On the Origin of Species* ensured that extinction became central not only to evolutionary thought but also to the ecological sciences.

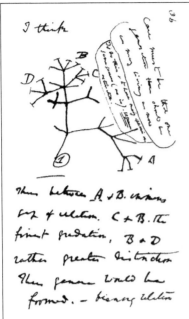

I think

Central to Darwin's conjecture about evolutionary change is the role of death and extinction in driving the transmutation of species. In Chapter VI of the Origin of Species, Darwin wrote:

"Hence, if we look at each species as descended from some other unknown form, *both the parent and all the transitional varieties will generally have been exterminated* by the very process of formation and perfection of the new form." [emphasis, mine]

What is evident in Darwin's mature thought is apparent even in this sketch (from a July 1837 entry in Darwin's notebook) that extinction could be of two forms: extinction of a **terminal** form where members of a population go extinct without leaving descendants, or extinction can be **phyletic**, that is, parental and transitional forms go extinct in the process of decent.

Image: Charles Darwin's 1837 sketch from his First Notebook on Transmutation of Species (1837). This work is in the public domain in the United States because it was published (or registered with the U.S. Copyright Office) before January 1, 1927.

Charles Darwin was the paradigmatic conceptual thinker: that is, he was a scholar who employed simple diagrams both to stimulate his own powers of reflection and to encapsulate the state of his thinking about theoretical issues. One of the most celebrated of his diagrams – no more than a doodle, really – is from about July 1837 when Darwin jotted down the first crude version of an evolutionary tree. This drawing appears in his "transmutation" notebooks, those being the journals in which he developed the first outline of his evolutionary thought. In the drawing, the ancestral forms of life are placed at the bottom of the tree; their descendants emerge at different points along the skinny trunk. The diagram thus depicts, in the roughest of forms, that most exquisite of thoughts: *all living things are related.* Above and to the left of the drawing Darwin wrote, enigmatically, yet perhaps appropriately, "I think."

One aspect of this famous diagram – and one that can be easily overlooked – is that it contains an important allusion to the *role of extinction* in Darwin's thinking. In fact, this explicit reference to extinction is recorded in the text accompanying the sketch. There Darwin wrote:

I think. . . case must be that one generation should have as many living as now. To do this and to have as many species in same genus (as is) requires *extinction*. Thus between A + B the immense gap of relation. C + B the finest gradation. B + D rather greater distinction. Thus genera would be formed. – Bearing relation to ancient types with several *extinct* forms [emphasis is mine].

Cryptic though these remarks are, I have reproduced them so that they may encourage you, reader, to employ your own journals, as did Darwin: to jot down drafts of your thoughts, and to diagram them freshly evulsed from your laboring mind. This sketch is important in that it discloses that the thought of extinction was a component of Darwin's thinking of evolutionary change from early in his reflections on the topic. By the time that the idea of evolution is fully fleshed out in

On the Origin of Species by Means of Natural Selection, the words "extinct" and "extinction" occur one hundred and eighty two (182) times. To put this number in perspective, it occurs twice as frequently as "struggle" – a key word in Darwin's lexicon, which occurring ninety-six (96) times – and about half as often as that other central Darwinian term, "selection," which occurs four hundred and twenty-one (421) times. When in 1859, Darwin wrote that "no one I think can have marveled more at the extinction of species, than I have done" (Chapter X, *Origin*) we can believe him: it's quite clear that extinction was not a peripheral matter in Darwin's understanding of the workings of the natural world, it is pivotal in his thinking.

By the time Darwin made the concept central to his evolutionary theorizing, the notion that the extinction of species might shape the history of life, while not exactly new, was still relatively innovative. A recognizably modern concept of extinction had been introduced into geobiology, as we mentioned above, by the French naturalist George Cuvier in the late eighteenth century. Cuvier has a number of claims to fame. One of them was the proposal that the history of the planet had been shaped by a number of catastrophic revolutions, although this theory of "catastrophism" sounded a little too close to the sort of biblical catastrophism that secular biologists and geologists prefer to avoid. However, Cuvier's thinking was not inspired by religious thought. Though Cuvier's promotion of a catastrophism was supplanted, for a while at least, by the more moderate claims of uniform and steady change in the history of life (uniformitarianism, to give the concept of slow but steady change its name), he, nonetheless, had been vital in making the very concept of extinction scientifically acceptable.

Cuvier's extinction theory was later championed by geologist Charles Lyell (1797–1875), though there were some important departures in Lyell's understanding of the process from Cuvier's presentation of the topic. An important departure from Cuvier in Lyell's account of extinction is that Lyell took the long view: extinction, when it occurs, is a slow process: no mass extinction of species occurred [3]. What is important for us is that Lyell's work was especially influential for Darwin [3]. Like Lyell, Darwin accepted that extinction in the past had occurred, and, agreeing with Lyell, he thought that, for the most part, extinction proceeds very slowly. As Darwin wrote in 1859,

> There is reason to believe that the complete extinction of the species of a group is generally a slower process than their production. . . [4]

That being said, Darwin also noted that. . . .

> . . .In some cases, however, the extermination of whole groups of beings, as of ammonites towards the close of the secondary period, has been wonderfully sudden [4].

The Victorian view of extinction, exemplified by Darwin, as being slow and progressive (in the sense of a senescent species giving way to a superior one) accorded well, as a number of cultural historians have noted, with the spirit of the time. However, this view of extinction as occurring at a stately pace has given way, in the later twentieth and early twenty-first centuries, to an appreciation of the occurrence, alongside a background rate of extinction, of rapid extinction events. The suddenness, and vast scale of some extinction events and the simultaneity of the loss across many taxa, has become one the issues of our time.

The distinction between a slower background rate of extinction and the greatly elevated rates that occur during "mass extinction" events is now well established in the scientific communities relevant for the study of the history of life (evolution, paleontology, conservation biology, biodiversity studies, and so on). The distinction, as we discuss below, is crucial in evaluating the significance of observations concerning contemporary extinctions and projections about the future.

14.2   So What, Exactly, Is Extinction?

Darwin's 1837 sketch of a rudimentary evolutionary tree is illuminating not only because he posits the fundamental unity of life, but because in it he also infers a role for extinction in evolutionary processes. This, as we have seen, was made clear by the handwritten text that accompanies Darwin's diagram. An examination of the drawing reveals not only a role of extinction (see a reproduction of the sketch on a previous page), but also conjectures about how extinction may occur. Ancestral species "1," in Darwin's sketch is placed at the base of the tree, and has radiated into several species (A-D), some of which are more related than others (as indicated by the distances between the terminal twigs). Thus, diversification of species – the one becoming the many – may be the primary theme of the drawing. But, just as importantly, if more subtly, Darwin's conceptual case study presents hints at the role that extinction plays in the transmutation of an ancestral form. Species 1 is no more. Now, this form of extinction – where an entire lineage is subsumed into descendent populations – is not the most intuitive form of extinction; but it clearly has been a feature in the history of life. These days, it is referred to as "phyletic extinction" [5].

A classic example of phyletic extinction can be seen in the evolutionary sequence of the genus to which we humans belong (that is, *Homo*): although the taxonomy of this genus is complex, crudely we can say that *Homo erectus* (an ancestral species), evolved into *Homo sapiens* (contemporary humans). *Homo erectus* therefore is phyletically extinct [6]. But in addition to this phyletic transformation of an ancestral species through time, another form of extinction has occurred in the human branch on the tree of life. The hominin branch has had many offshoots, but most of them terminated in an extinction event of another sort.

> Extinction is a natural process and ultimately will be the fate of all species. Environmental scientists distinguish two types of extinction: **phyletic** and **terminal**. In a phyletic extinction, one species evolves into another, whereas a terminal extinction has occurred when a lineage leaves no descendants.

In contrast to the notion of phyletic extinction is the idea of "terminal extinction." In a terminal extinction, a species leaves behind no living descendants whatsoever. The lineage is simply gone. There is some debate among paleontologists about the relative frequency of phyletic extinction versus terminal extinction [7]. A plurality of specialist opinion nudges toward the opinion that terminal extinction is the more prevalent – in fact, phyletic extinction is often referred to in the literature of the topic as "pseudo-extinction," since the genetic lineage of a taxon is preserved, to some extent, in its descendants. What may be more important, from the perspective of this book, is that understanding the patterns and processes of terminal extinction events are key to appreciating contemporary species loss: that is, the sixth extinction denotes a large increase in terminal extinction events.

When most people – scientists and nonscientists alike – refer to "extinction" it is usually terminal extinction that they have in mind: when the population of a given species has been reduced to zero, leaving no descendants behind. Though terminal extinction may not have been depicted directly in the Darwin drawing of 1837, a reading of the 182 references to "extinct" or "extinction" in *On the Origin of Species* makes it clear that he understood the significance of these terminal events.

Extinction, then, refers to the disappearance of a species. It has occurred when a population size has been reduced to zero; the process of decline axiomatically results in a decrease in genetic diversity within a lineage [8].

14.3   What Are the Drivers of Extinction?

A range of both extrinsic and intrinsic factors drives terminal extinction.

Extrinsic factors include the loss or fragmentation and degradation of habitat, predation, competition, and direct exploitation of species. All of these factors (with the exception of direct exploitation of some species by humans) can occur naturally. But humans have certainly accelerated each factor. Additionally, humans influence the distribution of species, by introducing species outside their original geographical range either deliberately or inadvertently. A proportion of these introduced species can become invasive in the ecosystems of their new range. All of the factors are intensified by climatic changes in the anthropocene. This reality underscores the urgency of tackling climate change and biodiversity loss simultaneously and in a coordinated fashion.

Let us consider each of the primary factors driving extinction in turn.

1) There is a well-characterized *relationship between the spatial extent of habitat and the number of species* in the community. This is encapsulated in "species area curves." As the area of habitat increases, the number of species found in a habitat will increase. The reasons why diversity of species increases with area are various: these include the great probability of encountering heterogeneity (or ecological variability) of the habitat as area increases and in these circumstances there is an increased probability of encountering those rarer species that can persist with very refined environmental needs. In addition, a larger area simply provides more refugia for less able competitors [9]. As a consequence of this relationship, when habitat area is reduced, fragmented, and its quality undermined it decreases the likelihood of a population retaining a viable size. This is true whether the reduction of habitat is natural or influenced by human activity. This factor alone is regarded as the greatest driver of the extinction of species. A time lag may occur between the loss of habitat and the loss of species. This is because a population of a species may persist, sometimes for decades, but the reproductive rates are not at a replacement level and the population is thus dwindling over time. This is referred to an "extinction debt" or as "latent extinction." It is worth mentioning that the geographic borders of a habitat may remain the same even as that habit becomes disturbed and degraded. Though geographically, the habitat remains as extensive as before, the loss of habitat quality reduces the effective area for some species (often rarer elements in the system) and can provide opportunities for the thriving of invasive species.

2) *Increased predation* pressure can lead to extinction. This often occurs when a predator expands it range. When formerly isolated populations are "discovered" by new predators, the results can be disastrous for extant populations that oftentimes evolved in the absence of these predators. Once again, the introduction of non-native species, this time predacious ones, can exacerbate pressures on a system. Increased pressure by novel predators on islands, in particular, can often be devastating to locally adapted communities.

3) *Intensified competition* can lead to extinction at least in local communities. That "exclusion" of poor competitors occurs in local communities is a central axiom of community ecology. When competitive exclusion occurs simultaneously across the entire range of a species then this can translate into the extirpation of an entire species [10]. The introduction of non-native species can be a means by which both the competitive and the predatory environment of native

organisms changes, with the result that some species are driven locally extinct [10]. Changes in competitive forces within communities can occur under conditions of both natural and anthropogenic climate change. When climatic conditions are altered, this translates into modified levels of a variety of environmental factors. Such changes can facilitate the spread of some species and the contraction in the range of others.

4) *Direct consumption* or exploitation by humans has led to extinction. This remains often the case with vulnerable marine mammal and fish species, often because vulnerable species are caught as bycatch in fisheries [11].

We can see from the above that most of the extrinsic factors driving extinction can be amplified by homogenizing global biota, making problems associated with invasive species a profound driver of local extinctions in natural systems.

In addition to these extrinsic factors driving extinction, extinction can occur for reasons *intrinsic* to that species. Because species possess intrinsic differences, some taxonomic groups seem to be more extinction prone than others.

The inherent factors driving the vulnerability of species include:

1) Inherent rarity is strongest inherent correlate of extinction. Species whose abundance and range size, are restricted are already more prone to extinction. The reason for this is that small populations are vulnerable to demographic variability. A few seasons of poor reproductive success can greatly diminish an already small population. Small populations are, as well, susceptible to problems associated with inbreeding. Adverse environmental conditions can have species-wide ramification if that species is already vulnerable through rarity.

2) *Body size* is also a correlate of extinction risk. Larger bodied creatures tend to maintain smaller population densities. This, in part, is related to ecological energetics – larger bodies being more energetically expensive to maintain [12].

3) *Specialized animals* are more vulnerable to extinction. This is because specialization decreases the range of resources upon which a species relies. In a changing environment, it means that if a key resource is threatened this has ramifications for all species relying upon that resource. In contrast, generalist species are often buffered to a considerable degree from extinction risk [13].

4) *Asexually reproducing species* adapt more slowly to changing environments and, at least some researchers have conjectured that, therefore, these organisms have a greater risk of extinction [14].

For an excellent and readable overview of the factors driving both natural and anthropogenic extinction see science writer David Quammen's *The Song of the Dodo: Island Biogeography in an Age of Extinctions* (1996) [15].

14.4 Identifying Mass Extinction

In contrast to the steady pace of background extinction, *mass* extinction refers to extinction events in which a significant proportion of the world's species go extinct over a brief period of time. The degree of elevation in extinction rate required for an event to qualify as a "mass extinction" and the duration over which such extinctions persist is a matter debate [16].

Five such mass extinction events are now recognized – despite the vagueness of the term "mass extinction," the existence of these five is uncontroversial (the existence of smaller events that may verge on similar rates are more controversial). The "big five" include events at the end of the Ordovician (439 million years ago (mya)), Devonian (367 mya), Permian (251 mya), Triassic

(208 mya), and Cretaceous (65 mya). The event at the end of the Cretaceous – the one that took out the dinosaurs – is perhaps best known; though this event resulted in decreases at the familial and generic levels (16% and 47%, respectively) this was not as severe as ones terminating other geological periods. The end Permian extinction is the mass extinction event that came closest to wiping out all of life on earth: approximately 95% of marine and 70% of terrestrial species went extinct rapidly. This amounts to about half of the taxonomic families that were present at the time.

Based upon the geological evidence, we can be confident that these extinction events are very different in kind from the processes associated with background extinction. Generally, the proposed causes of the major and minor mass extinction events in the history of the planet include asteroid collision, volcanism, global cooling, global warming, oceanic regression and anoxia where much of the world's oceans were depleted of dissolved oxygen creating unsurvivable conditions for more organisms. Of the major mass extinction events, asteroid collision is definitively associated with the end Cretaceous event. Volcanism contributed to events at the end Permian, Triassic, but perhaps was also to some extent a contributor to the end Cretaceous event. Global cooling is associated with some minor mass extinction events and also occurred at the end Cretaceous. Global warming contributed to the end Permian event; marine regression occurred at the end Triassic. Anoxia/transgression is a contributory factor to most mass extinction events (except end Triassic and end Cretaceous). As we can see from this, briefest of accounts, generally several factors interact to produce each and every mass extinction event.

14.5   The Difficulty of Declaring Extinction in an Era of Mass Extinction

We live in an age in which a definitive confirmation of a species' demise remains a relatively rare event, and, simultaneously, one in which the catastrophic acceleration of species extinction has been quite credibly claimed [17]. These two assertions – that is, the rarity of confirming an extinction and the apparent advent of a contemporary mass extinction event – are not precisely contradictory thoughts, though they do seem to be in tension. We will discuss below where the data supporting both of these claims emerge, but for now, we can simply note that the great caution exercised by scientifically informed organizations such as the International Union for Conservation of Nature (IUCN) in affirming the extinction of individual species might, inadvertently, make it difficult for some to comprehend the immensity of the contemporary "biodiversity crisis." (For an account of the difficulty of shaping public perception of ongoing extinctions see [18].)

A reminder from Chapter 13: **Biodiversity** is the variety of living forms – from genes to ecosystems – along with the wide diversity of processes that maintain the patterns of life on the planet.

The concepts required for understanding both the decline of biodiversity and its conservation seem to be relatively poorly understood by nonspecialists and the issue of extinction is not always regarded as a priority quite on the scale of the climate crisis (which now has registered as a concern for many people) [19]. To compound the problem, members of the general population who express alarm over the drivers of biodiversity decline – for example, land use change and habitat loss – are often from different sectors of society (different ages groups, political leanings, educational levels, and so on) than those who regard climate change as an environmental priority [20]. Thus, these two significant global environmental challenges are often considered separate issues by many. This is problematic from a policy perspective, however, because it is becoming clear that action on climate change and biodiversity need to be aligned: tackling one problem should assist in remediating the other [21].

Despite the apparent confusion among the nonscientific public, a consensus is emerging among scientists and environmental policy makers confirming that a loss of biodiversity on the scale that the world is currently experiencing, puts in jeopardy many aspects of our livelihood. Biodiversity loss may have long-term implications for food security, health, and for our quality of life [22]. Species loss on a global scale may be catastrophic (or in the words of some recent commentators we may be facing a "ghastly future") [23]. This is a claim that we examine towards the end of the chapter.

14.5.1 Declaring a Species Extinct: The Endangered Species Act

In September 2021, the US Fish and Wildlife Service proposed removing 23 species from the list of threatened and endangered species maintained in compliance with the United States Endangered Species Act (ESA) [24]. This is the single largest simultaneous elimination of species from the list for reasons of the assumed extinction of species since this list was first formulated.

The ESA is the premier legislative tool in the United States for averting species extinction and directing federal conservation efforts. Once it was signed into law in 1973, the act superseded several other pieces of federal conservation legislation (these include the Endangered Species Preservation Act of 1966 and the Endangered Species Conservation Act of 1969). The enactment of the ESA was motivated by a prescient recognition that "various species of fish, wildlife, and plants in the United States have been rendered extinct as a consequence of economic growth and development" [25]. As part of the process of directing conservation efforts towards the protection of threatened and endangered species, the federal agencies designated to oversee the implementation of the ESA (that is, the previously mentioned US Fish and Wildlife Service and the Commerce Department's National Marine Fisheries Service) regularly assess the status of listed species that are considered vulnerable across the globe. Currently, there are 2361 listed species [26]. The basis for removing a species from the list (a processes referred to as "delisting") is determined by the availability of information on the species' population and its ecological vulnerability. Delisting also takes into account legal and social factors, such as the existence of other legal provisions protecting the species, and the economic incentives and the social pressures that may sustain and conserve the species.

If, based upon all the available evidence, a species is deemed to have sufficiently recovered to the extent that it no longer requires federal protection, it can be removed from the list. Since the enactment of the ESA a total of 54 species have been delisted *due to recovery*; a further 56 species have been "downlisted" from endangered, the highest category of extinction risk, to threatened (a lesser ranking of risk) [24]. However, the more recent proposal *to remove* twenty three species was based not upon success of recovery, but rather upon a determination that these species are now considered extinct – no recovery can be expected.

Among the species removed from the federal list in this latest round are Bachman's warbler (last seen in 1988), the Flat pigtoe mussel (last seen in 1984), the Large Kauai thrush (last seen in 1987), *Scioto madtom* (a fish, last seen in 1957), and, controversially (since there had been some putative, though ultimately unconvincing, sightings in recent years), the Ivory-billed woodpecker (last confirmed sighting in 1944). On average, the delisted species had not been observed in about half a century, though at least one species, *Phyllostegia glabra var. lanaiensis*, an endemic Hawaiian species in the mint family (listed in 1991) had not been seen in over a century. Indeed, many species placed upon conservation lists since the 1960s are presumed to have been extinct before they were even listed (about 71 species in all were last seen before protections of the ESA were enacted) [27].

This is not to say that the ESA has failed these 23 species: rather, it is a reflection of two factors. Firstly, many of the delisted species, as we have noted, were likely extinct even before the ESA came into effect and therefore the protections afforded by the ESA could never have availed these species. Secondly, the fact that many of these species had remained on the list for decades even though they were probably already extinct, serves as a reminder that *the process of declaring a species extinct is undertaken with considerable caution.* To underscore the degree of deliberation taken in declaring an extinction, one should consider the fact that, previously, under the ESA, only four listed species had been confirmed extinct with another 22 considered possibly extinct following protection [27]. In the long run, since the goal of the ESA is primarily to *prevent the extinction of listed species,* the legal mechanisms for delisting are arduous by design [28]. Conservation organizations try to avoid what is known as the Romeo Error, that is a situation where by a declaration that a species is extinct when it is not, thereby withdrawing potential protection from that endangered species, and thus contributing to its subsequent extinction [29, 30]. It should not be surprising that delisting is therefore an inherently conservative process.

14.5.2 Declaring a Species Extinct: The International Union for Conservation of Nature's Red List of Threatened Species

International Union for Conservation of Nature: Categories of Extinction Risk

Category	Explanation of Category
Extinct (EX)	no longer extant
Extinct in the Wild (EW)	found only in captivity of under cultivation
Critically Endangered (CR)	species threatened with global extinction.
Endangered (EN)	species threatened with global extinction.
Vulnerable (VU)	species threatened with global extinction.
Near Threatened (NT)	species evaluated with a lower risk of extinction.
Conservation Dependent (CD)	species evaluated with a lower risk of extinction.
Least Concern (LC)	species evaluated with a lower risk of extinction.
Data Deficient (DD)	insufficient information for a proper assessment of conservation status to be made
Not Evaluated (NE)	not yet assessed to determining conservation status

Source: Pho sai / Public Domain CC BY 2.5.

Maintaining lists of species considered vulnerable to extinction, such as the ESAs threatened and endangered species list, has helped to focus global conservation strategies in recent decades.

The ESA's list is, however, just one such conservation list. Another significant list is *The Red List of Threatened Species* (sometimes referred to as a Barometer of Life), compiled and sustained by the International Union for Conservation of Nature (IUCN). Founded in 1948, the IUCN is an international organization that now comprises over a thousand civic organizations and governmental members that are committed to advocating for the conservation of nature. Many thousands of independent scientists contribute to the assessment of the multitude of species considered by the Red List.

The Red List was established in 1964, and it provides comprehensive information of the extinction risk status of animal, fungal, and plant species. The assessment in this case is performed on a global scale and reports on the status of species are designed to inform conservation action in the international arena. The IUCN also provides the tools necessary for the creations of regional red lists [31]. Listed species are assigned to one in a range of categories: Extinct, Extinct in the Wild, Critically Endangered, Endangered, Vulnerable, Near Threatened, Conservation Dependent, Least Concern, Data Deficient, and Not Evaluated [31]. The Red List is not, however, merely a rostering of those species of greatest concern; it is augmented by information (when it is available) about each species' biogeographical range, current population size and trends, habitat characteristics, threats, and is accompanied by suggestions regarding research needed for effective conservation. Finally, the list provides information on the economic uses of and trade in the species. In addition, each IUCN report suggests the conservation actions that are most likely needed to protect the species. The numerical goal of the Red List project ultimately is to assess 160 000 species. As of the end of 2022, 147 517 species have been assessed. To complete the assessments of the remaining 12 483 species or so, will require an even greater involvement of scientists willing to contribute expertise, and, if the task is to be achieved in a timely fashion, needs an acceleration in the rate of assessment. Even when the goal of 160 000 species is met, this, as we have seen in the previous chapter, is but a very small fraction of all the species described (approximately, 9%), and since a relatively small proportion of the total number of species are described and named, the list represents a vanishingly small proportion of all life on earth. The list, furthermore, is not especially representative of those taxa that are most cryptic (including certain invertebrates taxa, for example), even though species in these groups may be most vulnerable to extinction [32].

Name	EX	EW	Subtotal (EX+EW)
Total	897	79	976
GASTROPODA	267	14	281
AVES	159	5	164
MAGNOLIOPSIDA	103	30	133
MAMMALIA	85	2	87
ACTINOPTERYGII	78	10	88
INSECTA	58	1	59
AMPHIBIA	35	2	37
BIVALVIA	32	0	32
REPTILIA	32	2	34

All taxa with at least one species considered extinct (using IUCN summary statistic, Dec 2021.) Data from IUCN. EX, extinct; EW, extinct in the wild (See text for details)

The process of allocating species into different categories of vulnerability has been criticized as qualitative or subjective by some conservation biologists [33, 34]. Nevertheless, the IUCN Red List of Threatened Species applies a rule-based approach to risk assessment and this facilitates standardization. While a commitment to a rigorous process for assigning species to categories of vulnerability has not eliminated all criticisms of the potential subjectivity in the evaluation of species on the Red List, it does, at least, insure that the procedures are transparent. Furthermore, a set of rules for limiting the most obvious biases is made available by the IUCN and these are extensively discussed in the scientific community [35]. Ultimately, the assessment of species is undertaken by a group of taxonomic and ecological specialists from across the globe. Their proposed designations are then reviewed independently by other scientists before these recommendations are finalized and published.

So what does the analysis of Red List species reveal about species' vulnerability to extinction? Of the 147 517 species studied by the IUCN, 41 000 species (that is, 28% of all species) are threatened with extinction. With respect to specific evaluated taxa, 41% of amphibians, 27% of mammals, 34% of coniferous tree species, 13% of birds, 37% of sharks and rays, 33% of reef corals, 21% of reptiles, and 28% of selected crustaceans are threatened.

As is the case with the ESA's Threatened and Endangered species list, the Red List focuses, by design, on conservation action – thus, the success of Red Listing is ultimately measured by *averting* extinction. The extent that the Red List, and comparable projects, have successfully directed conservation action in achieving their objectives is revealed by recent reviews. For example, some careful analysis by Friederike C. Bolam and colleagues showed that even though ten bird and five mammal species from the list have become extinct since 1993, extinction rates could have been 2.9–4.2 times greater without conservation intervention [36].

Bolam and her colleagues summarized their results as follows:

> We found that conservation action prevented 21–32 bird and 7–16 mammal extinctions since 1993, and 9–18 bird and 2–7 mammal extinctions since 2010. Many remain highly threatened and may still become extinct. Considering that 10 bird and five mammal species did go extinct (or are strongly suspected to) since 1993, extinction rates would have been 2.9–4.2 times greater without conservation action. While policy commitments have fostered significant conservation achievements, future biodiversity action needs to be scaled up to avert additional extinctions. [36]

Despite the generally positive tone of the Red List project which focuses on conservation action rather than on loss through extinction, the list, nonetheless, includes a scrupulously curated list of those species assessed as being extinct (E) or extinct in the wild (EW), a designation for species where a residual population (or populations) may be maintained in captivity although wild populations of that species no longer exist. For the purposes of the Red List, the category of "Extinct" is defined as follows:

> A taxon is Extinct when there is no reasonable doubt that the last individual has died. A taxon is presumed extinct when exhaustive surveys in known and/or expected habitat, at appropriate times (diurnal, seasonal, annual), throughout its historic range have failed to record an individual. Surveys should be over a time frame appropriate to the taxon's life cycle and life form [35].

In contrast to the procedures associated with the ESA list of Threatened and Endangered species, there is *no process* for "delisting" a species from the Red List. If a species becomes extinct, it is retained on the list and merely re-assigned to the "extinct" category. In December 2020, for example, 31 species were officially declared newly Extinct according to the IUCN. These species included three Central American frog species; a further 22 frog species across Central and South America are now listed as Critically Endangered (Possibly Extinct) [37]. In December 2021, the dragonfly species St. Helena Darter (*Sympetrum dilatatum*) (last seen in 1963) was declared extinct, as was the plant *Licaria mexicana* (which has not been seen for over 80 years).

Guam Rail in the Cincinnati zoo. The Guam Rail (*Hypotaenidia owstoni*) illustrates how conservation listing can direct attention to critically endangered species. It had been declared Extinct in the Wild but thanks to a 35-year captive breeding program; the bird has now been reestablished on the neighboring Cocos Island. It is only the second bird in history to recover after being declared extinct in the wild [38]. *Source:* Greg Hume / Wikimedia Commons / CC BY-SA 3.0.

What does the Red List data reveal about extinction in the modern era?

As of the end of 2021, there were 897 species in the Red List categories extinct or extinct in the wild. These are species known to have gone extinct over the past 500 years or so (the year 1500 CE is the usual cut-off point for consideration by the IUCN). This list of species certified as either extinct or extinct in the wild is a mere 0.68% of all species on the Red List (only 0.62% when the category extinct is considered on its own), though this is widely acknowledged as an underestimate of the true extinction rate [39].

That the Red List underestimates extinction is understandable when one recognizes that the goal of the Red List, and of the ESA's and all comparable lists, is to focus primarily on actionable strategies directed at *what can be saved*, rather that upon *that which is irrevocably lost*. In keeping with this, the processes for categorizing species as extinct, though objective, *err on the side of caution*. To illustrate, in December 2021, six moth species that had been classified as extinct were reassigned to Critically Endangered (Presumed extinct) because it was unclear whether sufficient survey attempts to find them have taken place (IUCN staff, *personal communication*, 8 December 2021). So, it may be, and often is, the case that species in the category "critically endangered" are, in fact, already extinct. There are some instances where the species in one of the more vulnerable categories makes incremental improvements in its status: for example, a population of the Guam Rail (*Hypotaenidia owstoni*) has been established on the Cocos Island (in the northeastern Bay of Bengal). On the other hand, many species classified as critically endangered are often likely extinct. The Lost Shark (*Carcharhinus obsoletus*), for instance, is known only from three specimens from Borneo, and there has been no sightings of this species since 1934, despite extensive fish market surveys in recent years. Although it is classified as critically endangered it is highly likely that this species had gone extinct before much about it biology had been understood. Because several species that are classified as "critically endangered" are quite likely extinct, these species are often referred to as Critically Endangered (Possibly Extinct), or Critically Endangered (Possibly Extinct in the Wild). These tags identify species that are likely already extinct, though the extinction has not been confirmed. In some cases, as well, species that are classified as data deficient (that is, there is insufficient information for a proper assessment of conservation status to be made) are often of real conservation concern, and might be representatives of any of the Red List categories, including extinct species [40]. In total the number of species that are confirmed extinct, confirmed extinct in the wild, or presumed extinct stands at 2229.

The upshot of the foregoing accounts of the maintenance of the ESA list and the IUCN Red List is that both are valuable tools for directing conservation action, and yet, at the same time, these are *very conservative metrics of extinct rates.* As a consequence of an abundance of caution in declaring a species extinct, it is rare when extinction events come to the attention of the general public (as occurred by means of press releases in December 2020, in the case of the IUCN, and September 2021 for the ESA list). Without a doubt, the process by which a species is declared irrevocably lost from the Earth *should* be undertaken with great circumspection. It is also the case, as we will see in the next section, that the estimation of total extinction rates in the contemporary era are also made cautiously and conservatively. The conclusion, however, of these carefully undertaken appraisals of *cumulative extinction* apparently confirms the unsettling fact that extinction rates in the modern era have greatly exceeded the background rates of this irrevocable process.

14.6 Estimating Extinction Rates during the Sixth Extinction

Species do not endure forever; by most accounts over 99% of all species that have existed since the dawn of life have already gone extinct [7]. Claims that we are undergoing an acceleration of species' extinction during the Anthropocene – the Sixth Extinction – assess, on the one hand, contemporary rates of loss against those calculated for the previous mass extinction events, of which there have been five major ones [16]. However, these previous mass extinction events are detected by paleontologists by comparing the rates of loss over a relatively short geological duration against the ongoing background rate that occurred over the broader sweep of geological history. In evaluating claims regarding the Anthropocene extinction, biologists, therefore, must also ask, the degree to which contemporary rates are elevated compared to the *long-term background rates of extinction.* As we shall see, there is some debate about how best to conduct the comparison, but most recent estimates indicate that, very conservatively calculated, extinction is occurring in recent decades 100 times faster than the background rate (though the true rate may be 1000 times faster) [41].

Background rate of extinction accounts for 90% of all extinctions; mass extinction events may take out a significant proportion of biota extant at the time in which they occur, but since these event proceed over a relatively short duration, mass extinctions account for but a modest proportion of overall extinction [42]. Determining a long-term background rate of extinction against which the significance of contemporary extinction rates can be assessed, requires that scientists have access to a long-term record of extinction and requires inspection of very extensive fossil collections [43]. Basing calculations of a species' lifespan upon the geological record must be undertaken very meticulously; the procedure requires a very careful implementation of the species concept, since the characteristics of the living organism are not available when examining fossilized remains. This means that the paleontologist often operates with a broader definition of individual species than are available to biologists working with extant species – this can affect calculation of extinction rates. There is, additionally, another problem to contend with: that is, the notorious spottiness of the geological preservation of organisms and the biases that this introduces into estimates of extinction rates. The fossil record is disproportionately skewed in favor of preserving marine invertebrates, and whereas the taxonomic resolution of the fossil data can be determined, in some instances, at the species level, it is often the case that diversity in the geological record is estimated at higher levels of taxonomic resolution such as genera, and even families [44]. Finally, the time resolution of the fossil record is often only precise to within thousands, or even tens of thousands, of years [42].

Diatoms are a type of microalgae that are widely found in the both aquatic environments and in soils throughout the world. Unlike many microbial taxa that are poorly preserved in the fossil record, diatoms leave an ample record. Because of their diversity, diatom speciation and extinction rates have been used to examine the response of the phytoplankton communities to climate change [45]. Such studies can be helpful in determining the implications of contemporary climate change for marine biota. *Source:* CSIRO / Wikimedia Commons / CC BY 3.0,https://commons.wikimedia.org/w/index.php?curid=354, last accessed on 14 July 2022.

Because of the sort of problems we have just discussed, calculating how long individual species endure, understandably, is a fraught business [46]. But some general observations can be made. Firstly, there is wide variability of species' lifespans within taxonomic families, and, furthermore, different taxonomic group possess unique suites of traits that correlate with rates of both speciation and extinction. Broadly stated: species of marine invertebrates can last from 5–10 million years; mammals, about 1 million years; diatoms, 8 million years [47]. These average durations needs to be interpreted with caution, as fossil records are skewed toward the most abundant, widespread, and geologically longest lasting species [42]. Even if a typical species lasts for over a million year, the span is trifling considering the long history of life (originating, as it did, approximately 3.6 billion years ago). Based upon the sort of painstaking work undertaken by paleontologists, and recognizing the caveats we have alluded to above, a background rate of extinction can be calculated.

Getting these background rates of extinction correct is clearly an important matter: not only is it of inherent scientific interest – these calculations of an average species lifespan form a complex and interesting puzzle – but it is also profoundly important since it is against these background extinction rates that contemporary losses are compared [48]. By performing these comparisons we can determine the extent of accelerating species loss during the Anthropocene.

The approach used to determine extinction rates by comparing rates during our recent era with the background rate is illustrated by the following example. There are approximately 10 000 bird species. Recent estimates calculate that about 1.3% have gone extinct since 1500 CE. This calculation draws upon the IUCN red list and other estimates. This is translated into about 26 extinctions per million species per year (or 26 E/MSY; "extinctions per million species-years"). This is significantly higher than the estimated background rate of about one E/MSY (one species extinct per million species-years) before human impacts. This is a very conservative estimate: avian conservation biologists suggest the true rate could rise to 1500 E/MSY by the century's end. If this prediction turns out to be correct, it represents a truly calamitous loss of bird species [49].

Another example fortifies the concerns provoked by the bird data. In a study of a larger sample of vertebrate species, it was reported that 32% (that is, 8851 of 27 600 species) have decreasing population size and a constriction in their range. The data on vertebrates confirm that we are in the midst of a period of massive species loss. The conservation ecologist Gerardo Ceballos, along with his colleagues, describes this as "a biological annihilation" and argues that their analysis confirms that the Earth is going through a major extinction event, the sixth extinction [50].

Although the IUCN long-term data, which, in recent years, have been extensively employed to assess contemporary species loss, are partial (for example, of the 160 000, targeted for evaluation, invertebrate taxa are underrepresented) the signal is very clear. For all the groups systematically evaluated (e.g. mammals, birds, reptiles, amphibians, fishes, as well as those invertebrates such as snails, corals, and so on), the cumulative losses over recent centuries (and especially in recent decades) are massively above what would be expected without a human influence.

14.7 Is Biodiversity Loss Potentially Catastrophic?

It should be clear from the forgoing that the rate of species extinction in our era is greatly elevated above background rates. Recognizing that species are just one part of the complex system of life, conservation biologists and managers have evaluated levels of loss not only of species, but also of rare and vulnerable ecosystems. This is necessary because biodiversity loss can be measured at the level of species, or at the ecosystem or any other focal level within the hierarchy of global living systems. Since 2014 the IUCN has implemented a Red List for Ecosystems (RLE) risk assessment tool [51]. As of 2019, 2821 ecosystems in 100 countries have been assessed using the RLE protocol and more are being added rapidly [52]. Although it is still early in the RLE process, the expansion of biodiversity risk assessment tools beyond the species level is important and reveals the extent to which ecosystems are vulnerable to collapse. Loss at each level in the hierarchy of global biodiversity affects the whole, and yet, as we have seen, the behavior of slower, larger, levels within any hierarchy can be resilient in the face of disturbance.

We can use the same risk assessment tools discussed in an earlier chapter when asking about the potentially catastrophic consequences of species loss or ecosystem collapse. "Global catastrophic risk" in the model of Bostrom and Ćirković is evaluated in terms of its scope (the number of people that will be affected), severity (how badly these people would be affected), and its probability (how likely it is to occur) [53]. From this perspective, we can confirm that the sixth extinction event is truly global in extent, and the probability that the loss will continue into the future is very high. However, the severity of repercussions for human life and welfare is more difficult to assess. Indeed, the development of tools for determining the implications for health and economic harm is still in its infancy.

Even in the absence of precise conjectures about the human toll of incremental losses we can still gain reasonable insights into these consequences. Humans, undoubtedly, depend upon the role of biodiversity for the goods (food, fuel, fiber, medicines) and services (nutrient cycling, pollination, the regulation of ecosystem processes, and all the supporting serves derived from nature) it provides. From systems theory we have already established the fact that the loss of "parts" can, in some circumstances, be destabilizing to the functioning of the whole system. Obviously, there exists a point along a scale from zero loss to the collapse of all natural systems, when human civilization is no longer tenable in its current form; when the 1000-fold increase in species loss is considered alongside concomitant ecosystem degradation which continues to unfold over many decades or centuries, at what point are the conditions for humans not survivable? Short of total collapse, this question is currently impossible to answer [54]. However, the emerging body of research on the relationship between biodiversity and ecosystem functioning (BEF) continues to provide important insights into the functional roles that species play in ecosystems. This research also points to the manner in which diverse mixes of species act in a complementary fashion to provide resilience during periods of disturbance. This sort of research is especially important as disturbances due to climate change become ever more apparent.

Given the enormous levels of species extinction, the severity of habitat transformation, and the homogenizing of global ecosystems, adopting a precautionary approach would suggest that we conserve as many of the parts as possible. In his influential book *A Sand County Almanac* (1949), the conservation biologist Aldo Leopold wrote, "The key to intelligent tinkering is to keep all the parts." We are not tinkering with sufficient deliberation, and we are losing the parts.

14.8 Coda: A Role for Ethics

In addition to the delivery of goods and services, nature (and individual species, as well as companion organisms, i.e. our pets) can hold huge pyschospiritual value to humans. Nature is a source of inspiration and awe, and, in many traditions, nature exerts a powerful spiritual sway. From all of these perspectives, the human-driven extirpation of nature is an unfathomable catastrophe. It may seem easiest for policy makers to defend biodiversity primarily by appealing to its utility for us – its material and economic use. But nature also has intrinsic value – it has its own rights, a position developed by many influential ethicists [55].

One thing that I have found to be true is this: an admirable person wakes up in the morning with love, compassion and care on their mind. Love, compassion and care for their family, community, pets, the land and a whole variety of beings that do not serve our immediate needs. The annihilation of nature should offend our deepest sensibilities; perhaps reflecting on this alone can deflect us away from further disaster.

References

1 Callahan, B. (2013). Spring, on dream river.
2 Kolbert, E. (2014). *The Sixth Extinction: An Unnatural History*. A&C Black.
3 Hallam, A. (1998). Lyell's views on organic progression, evolution and extinction. *Geological Society, London, Special Publications* 143 (1): 133–136.
4 Darwin, C. (1996). *The Origin of Species: Oxford World's Classics*. Oxford University Press.
5 Flessa, K.W. (1979). *Extinction in Paleontology. Encyclopedia of Earth Science*. Berlin, Heidelberg: Springer.
6 Simpson, G.G. (1985). Extinction. *Proceedings of the American Philosophical Society* 129 (4): 407–416.
7 Benton, M.J. (2013). Causes and consequences of extinction. In: *The Princeton Guide to Evolution* (ed. J.B. Losos). Princeton University Press.
8 Rice, S.A. (2009). *Encyclopedia of Evolution*. Infobase Publishing.
9 Rosenzweig, M.L. (1995). *Species Diversity in Space and Time*. Cambridge University Press.
10 Bøhn, T., Amundsen, P.-A., and Sparrow, A. (2008). Competitive exclusion after invasion? *Biological Invasions* 10 (3): 359–368.

11 Kappel, C.V. (2005). Losing pieces of the puzzle: threats to marine, estuarine, and diadromous species. *Frontiers in Ecology and the Environment* 3 (5): 275–282.

12 Cardillo, M. (2003). Biological determinants of extinction risk: why are smaller species less vulnerable? *Animal Conservation* 6 (1): 63–69.

13 Gallagher, A.J. et al. (2015). Evolutionary theory as a tool for predicting extinction risk. *Trends in Ecology & Evolution* 30 (2): 61–65.

14 Frankham, R. (2005). Genetics and extinction. *Biological Conservation* 126 (2): 131–140.

15 Quammen, D. (2012). *The Song of the Dodo: Island Biogeography in an Age of Extinctions*. Random House.

16 Jablonski, D. (2005). Mass extinctions and macroevolution. *Paleobiology* 31 (2): 192–210.

17 Boakes, E.H., Rout, T.M., and Collen, B. (2015). Inferring species extinction: the use of sighting records. *Methods in Ecology and Evolution* 6 (6): 678–687.

18 Courchamp, F. et al. (2018). The paradoxical extinction of the most charismatic animals. *PLoS Biology* 16 (4): e2003997.

19 Rands, M.R. et al. (2010). Biodiversity conservation: challenges beyond 2010. *Science* 329 (5997): 1298–1303.

20 Skogen, K., Helland, H., and Kaltenborn, B. (2018). Concern about climate change, biodiversity loss, habitat degradation and landscape change: embedded in different packages of environmental concern? *Journal for Nature Conservation* 44: 12–20.

21 Pettorelli, N. et al. (2021). Time to integrate global climate change and biodiversity science-policy agendas. *Journal of Applied Ecology* 58 (11): 2384–2393.

22 Brondizio, E.S. et al. (2019). *Global Assessment Report on Biodiversity and Ecosystem Services of the Intergovernmental Science-Policy Platform on Biodiversity and Ecosystem Services*. IPBES Secretariat, Bonn, Germany.

23 Bradshaw, C.J. et al. (2021). Underestimating the challenges of avoiding a ghastly future. *Frontiers in Conservation Science* 1: 9.

24 US Fish & Wildlife Service (2021). US fish and wildlife service proposes delisting 23 species from endangered species act due to extinction. https://www.fws.gov/news/ShowNews.cfm?ref= u.s.-fish-and-wildlife-service-proposes-delisting-23-species-from-&_ID=37017 (accessed 16 November 2022).

25 USFW (US Fish & Wildlife) Service (1973). Endangered Species Act of 1973.

26 US Fish & Wildlife Service (2021). Summary of listed species listed populations and recovery plans. https://ecos.fws.gov/ecp/report/boxscore (accessed 16 November 2022).

27 Greenwald, N. et al. (2019). Extinction and the US endangered species act. *PeerJ* 7: e6803.

28 Doremus, H. and Pagel, J.E. (2001). Why listing may be forever: perspectives on delisting under the US endangered species act. *Conservation Biology* 15 (5): 1258–1268.

29 Roberts, D.L. and Fisher, M. (2020). Schrödinger's cat extinction paradox. *Oryx* 54 (2): 143–144.

30 Collar, N. (1998). *Extinction by Assumption; or, the Romeo Error on Cebu*. Wiley Online Library.

31 Brito, D. et al. (2010). How similar are national red lists and the IUCN red list? *Biological Conservation* 143 (5): 1154–1158.

32 Donaldson, M.R. et al. (2016). Taxonomic bias and international biodiversity conservation research. *Facets* 1: 105–113.

33 Harris, J.B.C. et al. (2012). Conserving imperiled species: a comparison of the IUCN Red List and US endangered species act. *Conservation Letters* 5 (1): 64–72.

34 Garcia-Macia, J., Perez, I., and Rodriguez-Caro, R.C. (2021). Biases in conservation: a regional analysis of Spanish vertebrates. *Journal for Nature Conservation* 64: 10.

35 International Union for Conservation of Nature and Natural Resources (IUCN) (2001). IUCN red list categories and criteria. https://www.iucnredlist.org/resources/categories-and-criteria (accessed 16 November 2022).

36 Bolam, F.C. et al. (2021). How many bird and mammal extinctions has recent conservation action prevented? *Conservation Letters* 14 (1): e12762.

37 International Union for Conservation of Nature and Natural Resources (IUCN) (2020). European bison recovering, 31 species declared Extinct – IUCN Red List. https://www.iucn.org/news/species/202012/european-bison-recovering-31-species-declared-extinct-iucn-red-list (accessed 16 November 2022).

38 Mann, M.E. and Gleick, P.H. (2015). Climate change and California drought in the 21st century. *Proceedings of the National Academy of Sciences of the United States of America* 112 (13): 3858–3859.

39 Regnier, C. et al. (2015). Extinction in a hyperdiverse endemic Hawaiian land snail family and implications for the underestimation of invertebrate extinction. *Conservation Biology* 29 (6): 1715–1723.

40 Luiz, O.J. et al. (2016). Predicting IUCN extinction risk categories for the world's data deficient groupers (Teleostei: Epinephelidae). *Conservation Letters* 9 (5): 342–350.

41 Ceballos, G. et al. (2015). Accelerated modern human–induced species losses: entering the sixth mass extinction. *Science Advances* 1 (5): e1400253.

42 Jablonski, D. (1994). Extinctions in the fossil record. *Philosophical Transactions of the Royal Society of London. Series B: Biological Sciences* 344 (1307): 11–17.

43 Foote, M. and Sepkoski, J.J. (1999). Absolute measures of the completeness of the fossil record. *Nature* 398 (6726): 415–417.

44 De Vos, J.M. et al. (2015). Estimating the normal background rate of species extinction. *Conservation Biology* 29 (2): 452–462.

45 Underwood, A., Chapman, M., and Connell, S. (2000). Observations in ecology: you can't make progress on processes without understanding the patterns. *Journal of Experimental Marine Biology and Ecology* 250 (1, 2): 97–115.

46 McKinney, M.L. (1997). Extinction vulnerability and selectivity: combining ecological and paleontological views. *Annual Review of Ecology and Systematics* 28 (1): 495–516.

47 May, R.M., Lawton, J.H., and Stork, N.E. (1995). Assessing extinction rates. In: *Extinction Rates* (ed. J.H. Lawton and R.M. May), 13–14. Oxford University Press.

48 Ceballos, G. et al. (2015). Accelerated modern human-induced species losses: entering the sixth mass extinction. *Science Advances* 1 (5): 5.

49 Pimm, S. et al. (2006). Human impacts on the rates of recent, present, and future bird extinctions. *Proceedings of the National Academy of Sciences* 103 (29): 10941–10946.

50 Ceballos, G., Ehrlich, P.R., and Dirzo, R. (2017). Biological annihilation via the ongoing sixth mass extinction signaled by vertebrate population losses and declines. *Proceedings of the National Academy of Sciences of the United States of America* 114 (30): E6089–E6096.

51 Keith, D.A. et al. (2015). The IUCN Red List of ecosystems: motivations, challenges, and applications. *Conservation Letters* 8 (3): 214–226.

52 Bland, L.M., Nicholson, E., Miller, R.M. et al. (2019). Impacts of the IUCN Red List of ecosystems on conservation policy and practice. *Conservation Letters* 12 (5): e12666.

53 Bostrom, N. and Cirkovic, M.M. (2011). *Global Catastrophic Risks*. Oxford University Press.

54 Naeem, S. (2002). Ecosystem consequences of biodiversity loss: the evolution of a paradigm. *Ecology* 83 (6): 1537–1552.

55 Naess, A. (2009). *The Ecology of Wisdom: Writings by Arne Naess*. Catapult.

Section Six

Conceiving a Future

15

Conceiving a Future: The Need for Interdisciplinarity

Interviewer: "What is your idea of happiness?" Alfred Hitchcock: "A clear horizon."
Excerpt from CBC's interview "A Talk with Alfred Hitchcock" (1964)

One of my environmental teachers advised me not to make predictions about events that may happen before I retire or die. That way, he assured me, I would not have to endure the disgrace of being wrong. Yet, in this penultimate chapter, I focus primarily on the future, both the immediate and the longer term. By means of the admittedly indeterminate laws of age-specific mortality, it is certain that I will experience less future than past; gauged by those same laws, I suspect that most of my readers will persist longer than I will (go ahead, Google me, I am probably already dead). Acknowledging that my mortal clock is ticking, and that the stakes of disgrace are somewhat low for me, I still regard any act of prognostication as a grave matter. I, therefore, make my remarks about the future in this section of the book with trepidation. Where I seem pessimistic, I regret distressing the reader who has valiantly made it to the end of this book in the hope of a happy conclusion. Where I seem optimistic, I suspect some will think me naïve.

TThe future is often viewed as a horizon: the line beyond which our ability to view anything has been limited. On the horizon of the landscape, the eye can take us no further. Yet, we can still develop a picture of what lies beyond. We do so by taking cues from the topography between where we are standing and the ultimate limits of the distant horizon. The more of the landscape we can take in, the more detailed our projections can be.

The study of the horizon by "futurists" is often founded on the idea that there is no single future. Futurists often imagine several possible prospects. A wide variety of tools exists to aid in this analysis, each of which depends upon how much of the landscape can be taken in. These tools include time series analysis, emerging issues analysis, the Delphi method of surveying panels of experts, simulation and computer modeling, and horizon analysis.

Image: Clouds over the grove (1878) Ivan Shishkin.

A Primer on Human Impacts on the Environment: The Conceptual Approach, First Edition. Liam Heneghan.
© 2023 John Wiley & Sons Ltd. Published 2023 by John Wiley & Sons Ltd.

Age specific mortality refers to the death rate associated with a specific age cohort within a population. The mortality rate is calculated by taking the total number of deaths per year at a specific age and dividing this by the number of living persons at that age.

Since these concluding chapters reflect upon the future, let us first say a few words about the general nature of future events. Undoubtedly, a time asymmetry exists whereby the past, though not always fully knowable, is, in an important sense, fixed or set, whereas the precise nature of the future is unknown and, therefore, is seemingly open. This openness of the future can be a source of comfort to us: better times, we assure ourselves, are surely ahead. Radically unsettled though the future may appear to be, nonetheless, it is possible to get a glimpse beyond the present horizon. The sciences play a role in making such predictions. Since our planet is governed by a set of atmospheric, hydrological, geophysical and ecological laws – ones that we still are far from fully understanding – these can be used to extrapolate beyond the present moment. However, because planetary processes interact in complex ways, the laws of the environmental sciences cannot be fully deterministic. In other words, environmental systems have some of the characteristics of chaotic systems: small alterations to variables at any point in time can result in large changes to the future state of the system.

Despite difficulties in making very precise forecasts, we can, nonetheless, develop a set of scenarios for the future. The process of scenario building recognizes that with each policy decision we make, certain other possibilities are foreclosed while yet others become possible. This generates the prospect of *several* possible futures, and yet, for all of this, the options are not limitless. Another way of indicating this state of affairs is to say that some, though, importantly, not all, aspects of our planetary future have been purchased in advance. For example, since vast amounts of anthropogenically generated carbon dioxide have already been pumped into the atmosphere and myriad species and ecosystems have been lost, this means that several possibilities for the coming decades, and perhaps even for the centuries ahead, are no longer viable options. To put this even more concretely: a changed climate is a certainty, though how grievous are the implication for human flourishing is not; the species lost in recent decades are gone forever and we face the future without them, but truly catastrophic implications for human life of Earth are still not inevitable. Devastating outcomes become more likely as we approach putative tipping points in the behavior of planetary systems, although it is not always clear where these critical transition points lie [1, 2]. Our lack of knowledge concerning *nonlinear* planetary responses – that is, the apparently random-seeming behavior that emerges in response to breaching important system thresholds – should compel us to exercise great caution [3].

The proverb, *"It is the last straw that breaks the camel's back"* which illustrates that the notion of **system nonlinearity** is a deeply intuitive notion. Our poor camel is fine until that last straw is added, but then, having exceeded an important threshold, the implications are disastrous.

15.1 Looking Back Before Contemplating the Future

To recap before orienting ourselves toward the future – scenarios about which future (or futures) are investigated in the next chapter – seems appropriate. A preparation for any thinking about the future, either scientific or in our practical dealings with the world, depends to some extent upon a recollection of our experiences. Neuroscientists have revealed that the "prospective brain" uses memory when imagining possible future events. Indeed, the very same neural machinery used for thinking about the future is employed by the brain in remembering the past [4]. In the first section of this penultimate chapter, we therefore reflect upon the material presented in the previous 14 chapters. This, at the very least, should convince the reader that they now have the conceptual tools at their disposal to understand contemporary environment problems and to contemplate the future no matter how bracing that future may be.

Over the course of the previous 14 chapters, we considered a suite of concepts that will serve well in evaluating anthropogenic impacts on the global environments. Concepts were defined in an early chapter as those general ideas that serve to represent the typical properties of things. Aggregated into complex frameworks, concepts allow us to think in a strong and universal manner about the world that surrounds us – they allow us to see both the similarities and differences between entities within the world. A conceptual approach to environmental studies allows the practitioner to apply their knowledge to the wide variety of systems they are likely to encounter in their work lives. Conceptual frameworks can assist in structuring the investigation of both existing and emerging problems.

Foundational to most of the frameworks reviewed in the book is the idea of a system – that is, the recognition that things in the world interact with various other things to produce a universe of complex, modular, and hierarchically structured entities, whose parts are linked by processes operating on a variety of scales. In the systems view everything is connected, but only up to a point. Some things are more connected than are others. *Restrictions* to connection can be just as determinative for the behavior of complex entities as connections can be. Just as ripples that propagate from a stone thrown in a pond will peter out at the shoreline (and the whole world does not quake because of this small disturbance), interactions among species, and between species and their environment, can dampen as we broaden temporal and spatial scales. The science of food web ecology, for example, – a thriving subdiscipline of environmental research – has as its core take the assessment of the relative strength of connections [5].

Systems are bounded modular entities. Boundaries can be either physical (like skin) or dynamic (determined by processes). Limits are important in structuring the world that we experience, but they also determine the bounds of system behavior. The material restrictions of the global ecosystem are influential in shaping humans health, well-being, and economic life. Throughout this book, I have argued that systems thinking is invaluable for environmental thought and practice because it allows us to hitch environmental insights to those emanating from other disciplines that also regard themselves as systems sciences.

After reviewing a suite of systems approaches to operationalizing the vague-seeming idea of an environment – the environmental mandala, the notion of an animal umwelt, biosphere, Gaia, noösphere and so on – we arrived at the concept of the Anthropocene, the proposal that we have entered a new geological epoch, one in which important planetary processes are now determined by the collective human influence.

15.2 The Anthropocene and the Noösphere

The term "Anthropocene" was introduced to the environmental sciences to denote that "human activities have. . . grown to become significant geological forces, for instance through land use changes, deforestation, and fossil fuel burning. . ." [6]. To briefly recap: in proposing the term, Paul Crutzen (and Eugene Stoermer, who is credited with an earlier use of the term), proposed the Anthropocene as a geophysical concept. It replaces the Holocene as way of stratigraphically naming our epoch. As a stratigraphical reference point, the term implies that the human influence now leaves its mark in the sedimentary deposits forming during our age: patterns of carbon deposits, distinctive radionucleotide accumulations, and a redistribution of nitrogenous compounds, for example. It is clear, however, that the Anthropocene is somewhat more than a geologically relevant term; it has cultural significance.

Because of the broad appeal of the Anthropocene concept, its study is the business of both natural scientists and humanists. The concept of the Anthropocene can be characterized as a "trading zone" (to use a term coined by historian of technology, Helmuth Trischler) for collaborating across diverse fields of thought and practice [7]. Ultimately, the Anthropocene is a *distinctive type* of idea; that is, it names *the idea that ideas change the globe* and will continue to do so. That the Anthropocene has, at its root, a recognition that *thought* is capable of transforming the planet can seem strange at first, particularly when we have conceived of the Anthropocene as measured by an outcome: for example, there is elevated carbon in the atmosphere. But it is clear that before the use by any human of a tool, was a glimmering of the tool's power; before the use of fire was the transfixing dream of the flame (see, Gaston Bachelard's book *The Psychoanalysis of Fire* (original French, 1938) for a highly unusual treatment of the effect of fire on the poetic mind [8]). The notion that many of the physical aspects of the world reflect human aspirations – that is, have been transformed by thought – is cogently developed in the *Masters of the Planet: The Search for Our Human Origins* (2012) by anthropologist Ian Tattersall [9].

The association between emergence of the Anthropocene and the *idea* of fire is an especially important one. During the course of human history, the appropriation of fire for human use was pivotal for human success. By means of the manipulation of fire – *control* is the verb often used in association with human exploitation of fire, but this seems like an exaggeration, as fire, to some extent, does what it likes – whole landscapes can be transformed, predators are deterred, food is cooked. That a direct line connects the control of fire and the emergence of the Anthropocene is articulated as follows by authors Will Steffen, Paul Crutzen, and John McNeill:

> The mastery of fire by our ancestors provided humankind with a powerful monopolistic tool unavailable to other species that put us firmly on the long path towards the Anthropocene. Remnants of charcoal from human hearths indicate that the first use of fire by our bipedal ancestors, belonging to the genus *Homo erectus*, occurred a couple of million years ago. Use of fire followed the earlier development of stone tool and weapon making, another major step in the trajectory of the human enterprise [10].

In chapter eight of this book, we considered some concepts that although they were proposed earlier than the Anthropocene are clearly related to it. For example, Antonio Stoppani, an Italian geologist, referred to ours as the era of the Anthropozoic. Another highly significant genealogical relative of the Anthropocene is that of the noösphere. Vladimir Vernadsky, the Ukrainian born mineralogist, was one of the three progenitors of that concept. He described the noösphere as an outcome of evolution toward increasing consciousness and thought. Unlike his collaborators, Pierre Teilhard de Chardin and Édouard Le Roy, who discussed the noösphere in a spiritual and philosophical context, Vernadsky saw the noösphere as a biophysical notion. As Vernadsky conceived it, with the emergence of the human being, the mind, as a collective force, becomes, for the first time, a transformative power on Earth. From this perspective, the noösphere is a product of the *biosphere*, a distinctive definition of which concept Vernadsky had previously developed with de Chardin to signify another collective force (that is, the biosphere understood as the aggregation of all living beings exerting an influence on global physical processes). From this perspective, mind evolved from out of the biosphere – humanity being subject to the same biological laws as the rest of living nature – but it now exerts a power independent of its origins. Vernadsky suggested that this power is best expressed through our scientific and technical know-how [11].

The more the noösphere concept took on spiritual overtones, especially through the work of de Chardin, the less appealing it became to some scientific thinkers. The immunologist and writer Peter Medawar (1915–1987), for example, described the greater part of de Chardin's ideas as "nonsense" [12]. In this, Medawar seemed to have been primarily commenting on the obscurity of de Chardin's prose and on the religious overtones in his thinking. Despite such criticisms, some other scientists were drawn to an interpretation of the noösphere as the universe becomes aware of itself. For example, in the introductory chapter of de Chardin's most widely read book *The Phenomenon of Man* (1955), English biologist Julian Huxley explicitly claimed, "Man discovers that he is nothing else than evolution become conscious of itself" [13]. The idea of the noösphere may be a controversial one, but the notion that human ideas can shape the world and determine its furture seems unassailable.

15.3 Ideas at the Interdisciplinary "Trading Zone"

The noösphere and the Anthropocene should correctly be retained as independent concepts, each attracting its own scholarship and each being applied to a distinct set of problems. However, when we evaluate humanity's prospects for averting the worst of anthropocenic instability, understanding the connection between the two concepts, Anthropocene and the noösphere, seems crucial. This is because activity within the noösphere – that is, movements emanating from the realm of ideas – will determine both our fate as well as the course of the evolution of the anthropocene at least for several generations.

Vernadsky, in at least some of his writings, seemed optimistic that this new sphere on the planet, the thinking sphere, could exert a positive influence. Not all environmental scientists agreed. In his *Fundamentals of Ecology* (1959), Eugene Odum was troubled by the assumption that the noösphere embeds a notion that "mankind is wise enough to understanding the results of all of his actions" [14]. Nevertheless, wise or not, humankind holds its fate in its own hands, or rather, we hold our fate in "mind" made visible through action.

If the noösphere has the capacity to shape the future of the Anthropocene, the study of this future will depend upon opening up exactly the sort of "trading zone" that Trischler identified, since reflecting upon the future is not the provenance of one discipline alone. One such trading zone is that which opens up between the environmental sciences and the humanities. The work of philosopher Michel Serres (1930–2019), who often reflected both upon environmental matters and upon the nature of human thought, is instructive from this vantage point. His work will serve as one example of the power of such collaboration in the trading zone. Serres writes explicitly "the task of philosophy is to anticipate the future" [15]. Like other writers on the Anthropocene (though this is not terminology Serres explicitly used), Serres sees the connection between our contemporary environmental problems and the deep human past as being mediated by thoughts of fire, or at the very least by the concept of heat. He wrote (echoing Will Steffen and colleagues, quoted above):

> As soon as human technology started using heat, vaporous mixtures expanded everywhere in all directions and at random; recent core sampling of glacial inlandis [*French, inland ice*] have been able to date the beginning of the bronze age almost to the year, thanks to traces of the first effluents emitted by archaic ovens in the Middle East which were dispersed everywhere and carried by snowfall to those high latitudes. Who would have thought that globalization started as early as our prehistory? [15]

Michel Serres (1930–2019) a thinker at the interdisciplinary trading post. Serres was a philosopher and environmental thinker whose later thought sought to enshrine a natural contract between humans and the rest of nature.

Credit: Michel Serres à l'Espace des sciences, 15 février 2011. Shared via creative commons (Wikipedia).

Serres traces the genesis of global environmental problems to the "thermal techniques that accelerated the rise of the local towards the global" [15]. More generally, he argued that our technological prowess leads to the creation of what he called "world-objects," that is, tools that scale up to the dimensions of the world, and in that category he would include ballistic missiles, earth orbiting satellites, the Internet, and nuclear waste. "Thought," Serres writes, "is nothing or tremendously powerful, it all depends" [16]. The genius of science and technology has been that it consistently tips the probability that "butterfly thought" is transformed into a "hurricane effect." To paraphrase: the noösphere – the collective expression of human cerebration – sometimes leaves no trace upon the world, and yet at other times it has anthropocenic impacts by means of a handful of pivotal interventions into the planetary processes. Big ideas matter. Novel thinking matters.

Serres was an exceptionally prolific writer who authored more than 60 books in his lifetime. The terms "noösphere" or "Anthropocene" are not used, as far as I can ascertain, across this monumental output. Nonetheless, a reader who comes to Serres' work from the environmental sciences will find much that is familiar in his thought: a natural history sensibility, the notion of environmental limits, a recognition of a planetary crisis, and a recognition that human innovation, especially in the form of technology (the production of "world objects"), can influence planetary affairs. There is also something conceptually akin to Vernadsky and de Chardin's "biosphere" in Serres idea of "Biogea," an entity that includes all of life, including that of the human realm [17]. The thrust of Serres later work makes the point that reflecting upon our ethical obligations within human communities will not be enough if we have forgotten the brute facts of the world. Through our current environmental crises, the brute world intrudes (in the form of resource needs and sinks for pollution) into contemporary politics, whether we like it or not (see the philosophical writer Peter Johnston's review of Serres last books for more details about Serres' as environmental philosopher on these points [18]).

That the conceptual frameworks employed by Serres seem lucid to the environmental scientist should not surprise us, since this is what one expects at the "trading zone." Lucid though it is, we should not expect, nor want, Serres simply to repeat what we find in a scientific text. The value of a greater interdisciplinary exchange in an evaluation of the future is that philosophy (as illustrated in this instance) will help develop a more vigorous understanding of how *thought* connects with the planet's processes. After acknowledging that through human action the world has become precarious, and "infinitely fragile," Serres analyzes our current situation by framing as a philosophical investigation of war and conflict. Since we have made of the world a fragile combatant, we will now need to impose on behalf of the injured Earth a legal contract that protects it in this conflict. This is what Serres is referring to in the title of his book *The Natural Contract* (translated from Le Contrat Naturel 1990).

It is not my intention here to investigate the impact of Serres's work in detail. After all, I selected him merely as an example of a writer who can inform us of how *ideas* (that is, movements in the noösphere) can affect the human relationship with the planet. The significance of this consideration of Serres is as follows. Human engagement with planetary processes reflects our distinctive human mental capacities. Because of this, many of our actions are influenced, on the one hand, by an ability to perceive the world as it is, and, on the other hand, by a desire to reconceive the world in the way we want it to be. Contemplating the human capacity for symbolic thought becomes key to understanding the future [9]. The tasks of interrogating the nature of symbolic thought, understanding human motivation, and scrutinizing the institutional context in which our capacity to change landscapes are determined is the provenance of many disciplines. Because of this, other thinkers and researchers in a wide variety of traditions must engage in the "trading zone." Along with environmental scientists, there are (to provide a truncated list) important philosophers, psychologists, historians, political scientists, policy scholars, legal thinkers, visual artists, and even novelists that can help us understanding the noösphere. To understand noöspheric dynamics the following interdisciplinary attributes seem key.

1) *A commensurable language*: each discipline of thought is likely to retain its own specific terminology; nonetheless, a shared lexicon will help coordinate the disciplines. Though each discipline does not need to employ a systems theory derived from the same key sources, recognizing the importance of identifying entities and the interactions that connect them seems vital to an interdisciplinary dialogue. Understanding how the behavior of systems emerges from the dynamics of parts and whole can serve as an important point of departure for analysis

2) An *acknowledgement of a common set of problems*: there are myriad problems confronting the globe that can be rightly considered environmental problems. Issues confront us with respect to soil, water, air, and the biota of the planet. There are no environmental problems that will not benefit from interdisciplinary evaluation; however, most problems will be exacerbated by biodiversity loss and climate change. To put this another way, amelioration of global environmental change (that is, climate instability and biotic loss) will assist in improving with most local problems.

3) A shared belief that *the future is worth speculating about* and that certain outcomes are sufficiently undesirable that they should be actively avoided.

15.4 Coda

Global environmental problems transcend any one discipline. The immensity of such problems – the fact that they overwhelm our understanding and seem to render us powerless to act on a scale matched with their enormity – suggests that we will need vastly different approaches in solving them than we have hereto tried. They require the collective power of human thought, which over the years has been splintered into disciplinary baskets [19]. In recent decades transdisciplinary research ironically has become its own discipline [20, 21]. The reader of this book is advised to reflect upon the sort of training they will need to undertake in the coming years, and the sorts of coalitions and communities they need to build in order to guide us toward a livable future.

We turn in the last chapter to specific visions of the future. Although, we may not be able to peer over the horizon for any great distance, but there are mirages shimmering in the distance. These suggestive premonitions are all we have to work with; let us rally our energies to avoid our worst nightmares, and achieve a livable future.

References

 1 Scheffer, M. (2020). *Critical Transitions in Nature and Society*. Princeton University Press.
 2 Ritchie, P.D.L., Clarke, J.J., Cox, P.M., and Huntingford, C. (2021). Overshooting tipping point thresholds in a changing climate. *Nature* 592 (7855): 517–523.
 3 Jackson, W. and Steingraber, S. (1999). *Protecting Public Health and the Environment: Implementing the Precautionary Principle*. Island Press.
 4 Schacter, D.L., Addis, D.R., and Buckner, R.L. (2007). Remembering the past to imagine the future: the prospective brain. *Nature Reviews Neuroscience* 8 (9): 657–661.
 5 Cohen, J.E. and Stephens, D.W. (1978). *Food Webs and Niche Space*. Princeton University Press.
 6 Crutzen, P.J. (2016). Geology of mankind. In: *A Pioneer on Atmospheric Chemistry and Climate Change in the Anthropocene* (ed. P.J. Crutzen), 211–215. Springer.
 7 Trischler, H. (2016). The Anthropocene. *NTM Zeitschrift für Geschichte der Wissenschaften, Technik und Medizin* 24 (3): 309–335.
 8 Bachelard, G. (1987). *The Psychoanalysis of Fire*. Beacon Press.
 9 Tattersall, I. (2012). *Masters of the Planet: The Search for our Human Origins*. St. Martin's Press.
 10 Steffen, W., Crutzen, P.J., and McNeill, J.R. (2007). The Anthropocene: are humans now overwhelming the great forces of nature. *AMBIO: A Journal of the Human Environment* 36 (8): 614–621.
 11 Pitt, D. and Samson, P.R. (2012). *The Biosphere and Noosphere Reader: Global Environment, Society and Change*. Routledge.
 12 Medawar, P.B. (2021). *The Art of the Soluble*. Routledge.
 13 De Chardin, P.T. (2018). *The Phenomenon of Man*. Lulu Press.
 14 Odum, E.P. (1959). *Fundamentals of Ecology*. WB Sanders.
 15 Serres, M. (2006). Revisiting the natural contract. *CTheory, Thousand Days of Theory*. https://journals.uvic.ca/index.php/ctheory/article/view/14482 (accessed 16 November 2022).
 16 Serres, M. (1995). *The Natural Contract*. University of Michigan Press.
 17 Serres, M. (2015). *Biogea*. University of Minnesota Press.
 18 Johnson, P. (2021). The inclusive philosophy of Michel Serres for our time of crisis. *Environmental Humanities* 13 (2): 459–469.
 19 Smith, R. (1997). *The Norton History of the Human Sciences*. WW Norton & Company.
 20 Frodeman, R., Klein, J.T., and Pacheco, R.C.D.S. (2017). *The Oxford Handbook of Interdisciplinarity*. Oxford University Press.
 21 Bendito, A. and Barrios, E. (2016). Convergent agency: encouraging transdisciplinary approaches for effective climate change adaptation and disaster risk reduction. *International Journal of Disaster Risk Science* 7 (4): 430–435.

16

The Three Futures

Dear friend, the wolves have always eaten the sheep; are the sheep going to eat the wolves this time?

Elias Canetti, *Crowds and Power* 1960 [1]

"The future is in our hands," we are sometimes told. Ideas and action that emanate from the complex social-ecological systems in which we are all embedded will determine which of the three possible futures outlined in this chapter will come to pass. The individual seems at times to be almost powerless given the enormity of mechanisms that shape our world. And yet there are "moments of reversal" where the power of collective action asserts itself and can influence our fate. *Source:* Study of Three Hands Albrecht Durer (1494) (Albertina, Vienna, Austria). In Public Domain (per Wikiart).

In writing about the cultural history of the number three – a number that is assigned practical as well as almost mystical significance – the American classical philologist and linguist Emory Lease (1863–1931) argued that this number has a symbolic role for at least three reasons. The number is important in many primitive conceptions of the world, it was extensively incorporated into the writings of the Ancient Greeks (including those of Aristotle), and, at least for those in the Christian tradition, its symbolic role is influenced by the Holy Trinity. The number three also manifests in folkloric tales, nursery rhymes, and in practical matters (for instance, in traffic lights and as beginnings, middles, and ends in the structure of a good story).

A Primer on Human Impacts on the Environment: The Conceptual Approach, First Edition. Liam Heneghan.
© 2023 John Wiley & Sons Ltd. Published 2023 by John Wiley & Sons Ltd.

Presenting three scenarios for the future, as I shall do in this final chapter, is, besides, a handy tool for forecasting. The approach serves to reduce an almost infinite number of possible futures (many of which are indistinguishable from one another) into manageable categories. For a range of environmental global parameters, we can project these three possible futures: a worst case, a best case, and the most likely scenario.

16.1 The Worst-Case Scenario

The worst-case scenario is a possible future in which, even as the implications of severe environmental degradation become increasingly more obvious, little or no remediation of these problems is undertaken. This was one of the scenarios envisioned in the report to the Club of Rome, *The Limits to Growth* (first published in the 1970s) [2]. According to its conclusions, the growth of industrial society would exhaust the earth's capacity to sustain its burgeoning population, emit enormous amounts of waste, and result in a collapse during the twenty-first century.

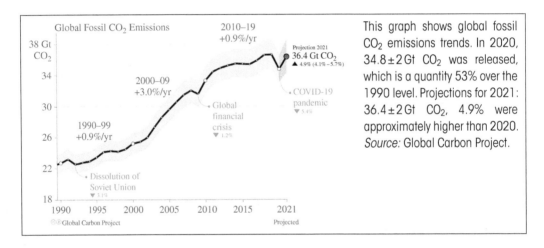

This graph shows global fossil CO_2 emissions trends. In 2020, 34.8 ± 2 Gt CO_2 was released, which is a quantity 53% over the 1990 level. Projections for 2021: 36.4 ± 2 Gt CO_2, 4.9% were approximately higher than 2020. *Source:* Global Carbon Project.

There remain *credible* scenarios in which the worst transpires. This is true not only with respect to environmental impacts but also for the important drivers of impact. For example, at present *human population growth*, an important driver of demand for scarce resources and of environmental degradation (though, it is worth stressing, that some sectors of the human population exert vastly greater pressure), is expected to stabilize early in the next century (and it is currently slowing, continuing a decades long global trend) [3]. However, population growth is highly sensitive to relatively small increases in total fertility rates. A small elevation of these rates would result in populations credibly exceeding 14 billion by 2100. After this time, growth could continuing unabated into the next century. This abundance would place an increasing strain on global ecological systems, and would make the sustainable conservation of resources harder to achieve [4].

A worst-case scenario for climate change emerges from the continued pumping of carbon into the atmosphere – a crucial driver of climate change. Although carbon emissions plummeted during the first year of the COVID-19 pandemic, emissions sharply rebounded in 2021. If climate policies are inadequately enacted, and current attention devoted to climate change peters out, then the elevation of global temperatures above preindustrial levels would climb to 4.1–4.8°C by 2100 and, thereafter, temperatures would continue to rise [5]. Even if business as usual proceeds and promises to achieve reductions in carbon emission fail to be met (as has been the case), then our planet becomes warmer and climatic systems will be more turbulent. This would be environmentally devastating and socially disastrous, making several parts of the Earth for all practical purposes uninhabitable (because of heat, drought and potential increase in the abundance of disease vectors). Consequently, many millions of people would be forced to migrate, exacerbating a global refugee crisis [6]. In the worst-case scenario, sea level would rise by over half a meter by the end of this century, resulting in severe flooding of coastal cities where a significant portion of the world's population live.

Negative interactions between climate change and biodiversity loss would be greatly exacerbated, as habitat characteristics across much of the globe are changed. We can think of this as representing a complete alteration in the condition of the environmental mandala (see Chapter 6). Undoubtedly, forested areas across the globe would face unprecedented fires because of a "moisture deficit" as a consequence of a changed climate [7]. These factors operating in combination would result in an unrecognizable world: security risks would be greatly elevated, severe economic repercussions would vastly decrease standards of living even in wealthier nations, and food systems which rely on biodiversity in very complex (and not fully understood ways) would be affected.

If the worst transpired, this could register as a collapse of civilization according to the terms of the definitions we explored in an earlier chapter: that is, there would be a rapid and significant loss of an established level of sociopolitical complexity [8]. Worst-case scenarios, though credible, remain somewhat unlikely. This is because it is already apparent that the management of environmental risk has emerged as an international policy priority – mitigation and adaptation strategies, even if insufficient, are likely to blunt some of the impacts, and will, more than likely, avert the very worst.

The planet Venus is suspected of having undergone a "runaway greenhouse effect." In the case of this planet, as the sun increased in luminosity early in the history of the solar system, concentrations of water vapor (from the planets early ocean) became elevated in the atmosphere, which, in turn, led to further increases in planetary temperature. Eventually, the feedback between the intensifying greenhouse effect, increased temperature, and more water vapor in the atmosphere resulted in the ocean boiling. The current temperature on Venus is approximately 450°C. Few climate scientists argue that a "runaway greenhouse effect" is likely on Earth, though this concern is not without its champions [9, 10].

Image from the Mariner 10 mission was managed by NASA's Jet Propulsion Laboratory. Data by JPL engineer Kevin M. Gill (2020) *Source:* https://photojournal.jpl.nasa.gov/jpeg/PIA23791.jpg/Author. NASA/JPL-Caltech NASA images are public domain.

None of the worst-case scenarios sketched above take into account the possible activation of tragic nonlinearities in the behavior of planetary processes. This is where planetary thresholds for climate, the loss of biodiversity integrity, and ocean acidification, to give several examples, are breached. If these system boundaries are ruptured, it could result in even more devastating implications for the social-ecological systems that were contemplated above [11, 12].

If the scenario above does not seem terrifying, it is worth reflecting that the number alone can impede the imagination. For this reason, imagining the very worst futures is often the task of science fiction. Good examples of the genre include the prescient novel by Kim Stanley Robinson, *The Ministry for the Future* (2020), and Octavia E. Butler's grim but important post-apocalyptic novel, *Parable of the Sower* (1993).

16.2 The Best-Case Scenario

Prospects for a *best-case scenario* emerge when a strong policy vision is implemented immediately to avert a suite of environmental disasters. Global action over the last century to curb the release of stratospheric ozone-depleting chemicals provides evidence that under the best of circumstances environmental problems can be addressed [13]. The sort of consensus action that was executed regarding ozone depletion seems harder to achieve for climate change and biodiversity loss [14]. In the case of climate change, for example, this would consist of achieving a transition to an economic system operating with zero-carbon emissions. Such a transformation in the economy can be imagined as requiring the adoption of a wartime level of response to environment problems [15].

> The best-case scenario will only come about through **transformative change**. Such change would involve a rapid decarbonization of fuel supplies, unprecedented commitments to energy efficiency, changes in the management of landscapes, and a rethinking of lifestyle choices.

Under the best-case scenario, carbon emission is almost immediately curbed by an unprecedentedly swift transformation of the energy economy. *Transformative change* at a global scale would result in limiting global warming to below 1.5°C above preindustrial levels. Limiting the temperature rise to this level has been the aspiration of policy makers since the signing of the Paris Agreement, the legally binding international treaty on climate change adopted in 2015 [5].

The window of opportunity for adopting such a transformative response is shrinking. During the writing of this book, several leading environmental advocates have warned that "it's now or never." For example, in discussing catastrophic climate change, a recent United Nation's report stated that in order to limit global warming to around 1.5°C "global greenhouse gas emissions would have to peak before 2025 at the latest, and be reduced by 43% by 2030." A commitment to abating environmental stressors at this scale admittedly now seems unlikely [16]. A less ambitious, yet nonetheless optimistic, scenario is one where carbon emissions are phased out by 2100. The earlier this occurs, the more likely it is that global temperature rise remains below the critical threshold of 2°C by the end of the century. This too would require concerted efforts on the part of all the major global economies and a commitment to climate adaptation.

> Achieving the best outcomes with respect to climate change and other environmental calamities requires not only **climate mitigation** (e.g. carbon reduction) but also **climate adaptation**. Adaptation requires building both social and physical structures that protect us from the expected effects of climate change.

Several ambitious plans exist for stanching the loss of biodiversity. These include setting aside 50% of the Earth for conservation purposes [17]. Setting aside half the Earth for nature is a bold goal. If achieved it would undoubtedly halt or, at least, reduce the rate of biodiversity loss (you will recall the importance of the relationship between habitat area and the maintaining species diversity). Business as usual in the world of biodiversity conservation and ecological restoration is slowing the loss of biodiversity somewhat, but improvement of conditions for biodiversity are not on a sufficient scale to achieve the outcomes needed without the implementation of transformative solutions [18].

Undoubtedly, striving to achieve best-case scenarios, especially when transformative change is invoked in the face of uncertainties, can be economically, socially, and politically costly to achieve. Since reactive rather than proactive policy – especially when policies are implemented by top-down, centralized, bureaucracies – tends to be the norm, therefore radical change may be slow in coming. On the other hand, an aspiration to achieve best possible outcomes can unleash creativity within local communities. There is already evidence that many communities around the globe are not waiting for top-down responses and are responding innovatively to the local circumstances in which they find themselves. This should strike us as being highly encouraging.

Just as imagining worst-case scenarios can be aided by reading dystopian fiction, conceiving brighter futures can be helped by engaging with utopian fictions and visions. Though rarely read these days, *The Story of Utopias* (1922) by American urbanist Lewis Mumford is a good starting point to orient yourself to at least the classical forms of this literature.

16.3   The Most Likely Outcome

The most likely outcome occurs when at least some mitigation and adaptation efforts are pursued, as is presently the case. Environmental systems are vastly complex and are chaotic in the sense that small variations in initiating conditions can have large effects. The further out we look, the less confident we can be; the more precautious we are now, the better the conditions of the future will be. Assuming that no catastrophic transitions occur in the near future – such catastrophes would make present day predictions moot – then we can envision the medium-term future of certain variables with reasonable conviction.

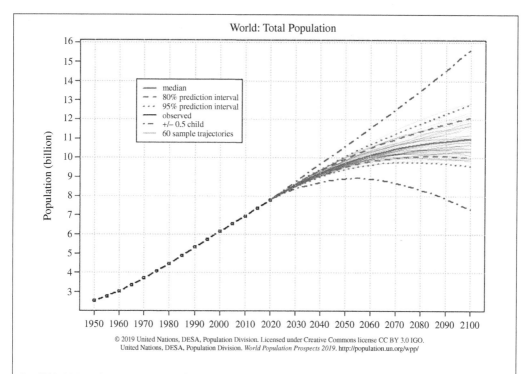

Total World Population Projection. *Source:* Reproduced from From UN Population Division, 2019 / Public Domain CC BY 3.0.

Population projections are based on an assessment of total fertility and life expectancy at birth. Small variation (+/−) in expected fertility in the coming decades could result in total populations growing to as many as 15 billion by the end of the century, or dropping to less than the current population (8 billion).

Under the *most likely scenario* presented in the most recent revision of the United Nation's *World Population Prospects,* the population is expected to reach 10.9 billion in 2100 [3]. This is a vast number, and at least some analysis suggests that this population level comes close to breaching the carrying capacity of the planet [4, 19, 20]. This was the conclusion of the monumental analysis by demographic ecologist Joel Cohen in his book *How Many People Can the Earth Support?* (1995). Cohen conclude as follows:

> Counting the highest estimate [of carrying capacity] when an author gave a range of estimates, and including all estimates given as a single number, the middle value or median of the estimates was 12 billion; counting the lowest estimate when an author gave a range, and the single number otherwise, the middle value, or median, of the estimates was 7.7 billion.

Feeding the world's population – which will likely rise to levels somewhere between the low and high estimates for a sustainable population – will certainly require a continued commitment to a model of highly industrialized and intensive agricultural. This requires immense quantities of pesticides and nitrogen fertilizer, both of which have negative ecological implications, and are produced in energetically costly ways [17]. This level of agricultural commitment also places profound pressures on global biodiversity. Agricultural dependence alone does not drive biodiversity loss, but the continued clearing of lands, and the removal of plant biomass to sustain human life, especially in ecologically vulnerable regions of the world, will compound the problems of extinction [21]. These problems will be amplified by climate change. In the absence of radical improvement in the efficiency and intensity of agriculture, sustaining food production for the human population will emerge as the greatest driver of future biodiversity loss [22].

With respect to climate change, the current aspiration of the international community is to keep warming "well below 2°C"; *ideally,* warming should be held to 1.5°C above preindustrial levels. Current reduction levels are not on track to reach the targets of the Paris Agreement, and taking into account current national reduction pledges on carbon emission warming by 2.5–2.9°C by the century's end can be realistically expected, though some other forecasts expect 2.8–3.2°C [23].

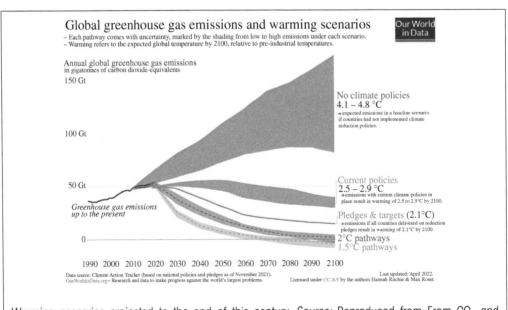

Warming scenarios projected to the end of this century. *Source:* Reproduced from From CO_2 and Greenhouse Gas Emissions, 2017 / Our world in data / Public Domain CC BY 4.0.

Under the most likely scenario, global ecosystems may very well endure without a catastrophic collapse well into the next century and conceivably beyond this. However, there will be continued erosion of general environmental quality. Environmental problems with soil, water, atmosphere, and biota on a local scale will create challenges that are costly to remediate. This will certainly result in a loss of some unique natural and cultural systems. Indigenous cultures and impoverished communities will remain especially vulnerable. Increasingly devastating climatic events will include periods of extreme heat and drought in some regions. It is also a certainty that environmental impacts will be graver in poorer nations – a continuation of long-enduring trends in environmental injustice [5]. Although the human and economic costs will be greater in poorer nations, even in wealthy nations the cost of climate mitigation and replacing diminished ecosystem services will be extremely expensive. We are, almost without a doubt, facing a hotter, biotically impoverished, and more unjust future.

Neither crystal ball nor data-rich forecasting can allow us to see very far beyond the next century. When cartographers of former years were confronted with unknown terrain, such lands were labeled Terra Incognita on the map. We are collectively embarking upon that part of a voyage where we are leaving the terrain that had been mapped out for us and are now sailing towards Terra Incognita. The future is always obscured by clouds, but the clouds of the Anthropocene are billowing, and, at least to some extent, are uncannily of our own making.

16.4   Fat-Tailed Risk

There is an emerging consensus among environmental economists and policy advocates that the probability distribution of key environmental impacts are "fat-tailed" risks. Under the assumptions of a normal probability distribution (that is, the familiar "bell-shaped" curve) extreme consequences have a probability of 2% (2% being the area under the curve at the edges of the normal distribution). Under the assumption of fat-tailed risks, catastrophe is much more probable. If climate change risks are indeed fat-tailed, it means that warming in excess of that predicted by average climate models will be more likely to occur [24]. According to some analysis there is now a 10% chance that large scale and probably catastrophic events will come to pass [25]. The medium-term future, under the assumptions of fat-tailed risk will be lethal for more people than is currently predicted, and the future becomes less endurable for most; this can be asserted with moderate confidence. Because of the nature of fat-tailed risk, a failure to strive for the best-case scenario flirts with catastrophe. Another way of stating this is as follows: if global environmental problems are truly fat-tailed risks, the most likely future converges with the conditions of the worst case scenario.

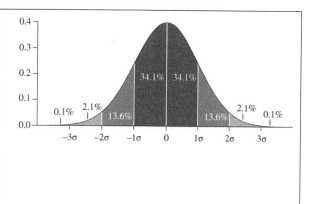

In this normal probability distribution curve, standard deviations from the mean are as depicted above. Each band in the diagram shows 1 standard deviation. If environmental repercussions followed a normal distribution, then the probably of worst-case outcomes would be low (around 2%). However, if environmental risk (especially climate risks) are "fat-tailed" as some have argued, then the risk of severe outcomes are as much as 10%. *Source:* M. W. Toews / Public Domain.

There is a tendency for environmental scientists to "err on the side of least drama" and disciplinary leaders are reluctant to seem pessimistic, but the reader of this book should not emerge feeling that a benign future will be easily achieved [26]. It certainly will not be easy; to achieve an affable future we may need a revolution in thought and action [27].

16.5 Revolutionary Transformations

There are perhaps good reasons for the skepticism that some academicians have about the crowd-sourced encyclopedia Wikipedia [28]. Despite ongoing debates about its reliability (though it is at least as reliable as the Encyclopedia Britannica), there have, nevertheless, been continued calls in recent years for academics to become more involved in writing technical articles and for editing in areas of their expertise [28, 29]. Although I have not cited it extensively as a reference in the text of this book (though I have reproduced a number of helpful graphs reproduced on that site), nonetheless, I regard it as a good point of departure as you continue in your reading about several of the concepts discussed in the book. There is, however, one article that I would especially like to bring to your attention, and that is the impressive page entitled *List of Revolutions and Rebellions* [30].

The Wikipedia page on revolutions and rebellions was created on 18 June 2006 at about three in the afternoon by a user called Sanmartin (Sanmartin is the Brazilian Wikipedian, José San Martin). The tally of revolutions and rebellions now stands at 1200, though the article is flagged as "incomplete." Incomplete though it may be, this page may constitute that handiest list of its kind since each of its 1200 named insurrections links to its own page. Some of these revolutionary moments were small affairs, while others were all encompassing. We undoubtedly live in the shadow of events that were both revolutionary and rare: the triad of the agricultural, urban, and industrial revolutions merely being the most widely discussed among them.

What exactly is a revolution (or rebellion) and is there need for a revolution in the face of the very real crises confronting us? This is certainly been a consistent call in environmental circles. Such calls are expressed in a desire for "a sustainability revolution" in such books as Andrés R. Edwards *The Sustainability Revolution: Portrait of a Paradigm Shift* (2005) [27]. More militant-sounding calls for a revolution are found in the writing of theorists such as Herbert Marcuse. In an influential essay *Ecology and Revolution* (1972) Marcuse, a philosopher and social theorist, insists that an authentic ecology "flows from a militant struggle for a socialist politics that must attack the system at its roots, both in the process of production and in the mutilated consciousness of individuals" [31].

Revolution is one of those words with a deep history (traced to a late fourteenth century French word referring to a celestial rotation about the earth) and is now used in a variety of contexts. In an innocuous register, revolution can mean a twist, turn or rotation, but it also has heftier social implications, for example, when it refers to change and upheaval. One of the definitions provided by the *Oxford English Dictionary* is the "overthrow of an established government or social order by those previously subject to it; forcible substitution of a new form of government" [32]. However, when I use the term in the context of discussion about the changes that may be required to bring the best-case scenarios into being, I have more of an anthropological definition in mind. When the anthropologist V. Gordon Childe described the urban revolution, he was at pains to insist that the word "revolution" does *not* necessarily denote a sudden violent catastrophe. Rather, he has in mind a culmination of a set of changes in the economic structure and social organization of communities that, in the case of the urban revolution (that is, the emergence of the world's first cities), was accompanied by population changes [33].

The 17 Sustainability goals promoted by the United Nations include:

(1) Zero Poverty; (2) Zero Hunger; (3) Good Health and Well-being; (4) Quality Education; (5) Gender Equality; (6) Clean Water and Sanitation; (7) Affordable and Clean Energy; (8) Decent Work and Economic Growth; (9) Industry, Innovation, and Infrastructure; (10) Reduced Inequality; (11) Sustainable Cities and Communities; (12) Responsible Consumption and Production; (13) Climate Action; (14) Life Below Water; (15) Life On Land; (16) Peace, Justice, and (17) Strong Institutions; Partnerships for the Goals.

The environmental revolution will surely need to be more rapid that envisioned by Childe, and it may not necessitate significant changes in world population (though fertility rates could plummet with universal access to education, with increases in quality of life, and with widespread availability of birth control). Rather, this revolution will be a revolution of *ideas* (that is, it is a movement from within the noösphere): it entails a conceptual transformation in economics and in the social-ecological sphere. Although each person on earth need not share an identical worldview, nonetheless, radical change will require a shared commitment to living on a safe, clean, and biodiverse planet in which environmental and social justice serve as bedrock principles. Although there have been numerous attempts to develop blueprints for our planetary future, the 17 sustainability goals proposed by the United Nations represent a comprehensive set of goals that many people will find motivating [34, 35]. They include measurable goals for achieving justice, equality, and environmental stability. Achieving such goals requires coordination at all levels of society from individuals, and in the corporate, industry, agricultural, and services sectors.

16.6 How to Hack the Future!

The type of "butterfly-wing" causality that Michel Serres described in his book *The Natural Contract* suggests that the world is shaped by rare innovations, the impact of which are not initially discernable. In the usual course of things, we will not be individually responsible for the creation of "world objects"; neither you nor I invented the equivalent of atomic bombs or the Internet. The chance that you will change the world by yourself is slim (though not, of course, impossible). Though this is undoubtedly true, nonetheless, human thought – that most unusual of butterfly wings – can be tremendously powerful. This is because thought, collectively expressed in the noösphere, has undoubtedly, albeit inadvertently, shaped the Anthropocene, the epoch in which we live.

Which of the three possible futures sketched above – the best case, the worst case, or the most likely case – will come to pass depends upon a level of coordinated, though variegated, thinking that has never occurred in history. We have changed the functioning of the planet in dangerous ways accidentally; can we now improve it intentionally? It is easy to feel pessimistic, and yet when we examine the tools at our disposal perhaps matters are not so grim. The British biologist Julian Huxley may have been correct when, in reflecting upon the noösphere, he wrote, "the universe is becoming conscious of itself, able to understand something of its past history and its possible future."

Can the collective consciousness of the universe – if that is how we chose to interpret the noösphere – deflect us from the most tragic consequences for humanity? In answering this question and pronouncing upon next steps, we can draw upon the same systems framework employed throughout this book. The following principles can be applied:

1) An axiom of general systems theory is that the behavior of the whole emerges from the interaction of the parts, while, reciprocally, the whole *constrains* the behavior of its constituting

elements. Furthermore, the tempo of the whole (and the spatial scale upon which it operates) is slower (and large) than of its parts (which are confined to smaller spatial scales).

2) Derived from this axiom we recognize that global processes of biogeochemistry and planetary governance will always proceed at a stately pace compared that which pertains to the scale on which we conduct the affairs of our ordinary lives.

3) For this reason, planetary systems tend to be *conservative* in the sense that the global behavior reflects an averaging of the behavior of the parts. The upper reaches of the hierarchical system act as a repository of crucial feedbacks (the system's "memory"). This memory ensures that anomalous variation in the behavior of elements in lower levels in the hierarchy is dampened out, or suppressed. This form of *vertical decoupling* ensures that upper levels are largely impervious to change initiated from below. The conservative tendency is referred to a systems' capacity to *remember*. System stability in the face of ephemeral perturbation depends upon this capacity [36].

4) But there *are* those occasions – and this is the critical point – when behavior initiated in lower levels can upset the whole. In both system and social science, these occasions, as we have alluded to above, are referred to as *revolutions*. In systems terms the capacity to "revolt" can spark critical change and this change can cascade up through the hierarchical levels and overwhelm conservative forces. In such a manner, the irreversible altering of large, slow, systems is achieved.

5) In his important (and very readable) account of the role of crowds and packs in political affairs *Crowds and Power* (1960) (*Masse und Macht*, in its original German) – a quote from which is the epigram for this chapter – the Bulgarian writer Elias Canetti describes several occasions when combined elements of the social system (the crowd) overwhelms the societal whole ("the sheep eat the wolves"). Canetti calls this the "essence of reversal." Many of the examples that he provides are local revolutionary moments and such reversals are not sustained for long durations (many of the over one thousand rebellions listed on the Wikipedia page mentioned above are short-termed). Thus, the crowd seizes power temporarily, but then the old order is restored, equanimity is achieved, and the parts return like frisky cattle to their yokes. Canetti's theory of reversal crowds is lucid from the perspective of general systems theory.

6) But reversals of a more durable nature can and do happen – the French and American Revolution (and the Enlightenment) being examples that are merely more conspicuous. Very few regions have not had their revolutionary moments. These are often crystallized around new thinking, adopt novel principles, and become the preoccupation of important movements. It is to a consideration of such movements that we turn in the final moments of this book, as we reflect on how we might cultivate transformative change.

We can extrapolate from these general principles to envision how collective thought and action can help avoid tragic consequences of environmental change for humanity. Social movements represent coalitions of organizations and activists that engage at local, national, and international levels and that advocate for ethical change based upon a set of core principles. Such movements often develop around eclectic aspects of a common framework. The *environmental movement* is one such diverse coalition. It advocates for the conservation of lands (including wilderness areas), the protection of biodiversity, for clean air and water, for improvements in environmental health (especially in urban areas), for fair access to green space, for food justice, for energy alternatives to fossil fuels, and for general sustainability in resource use. As often as not, diverse environmental campaigns advocate for alternative economic systems, for environmental justice, and for the implementation of antiracist policy. Many of the aims listed in this eclectic agenda intersect with other influential social justice movements (including, for example, civil rights and universal suffrage).

This diverse coalition is not fully coordinated in its immediate aims nor does it always agree about the methods used to achieve its ends. However, a full concordance may not be essential for transformative change. That being said, from a system point of view in order for the "parts" to coalesce to achieve change, there needs to be at the very least a concerted push in the same direction. How this coordination is best achieved goes beyond the immediate scope of this book but insight can be gleaned from both classical and contemporary political theorists (see the edited book by sociologist Christopher Rootes, *Environmental Movements: Local, National and Global* (2014) for some insights [37, 38].)

The prospect of an imminent extinction of humanity seems remote at least in the short-term, but global environmental problems – climate change, biodiversity loss, erosion of soil quality, the fundamental alteration of global biogeochemistry – are potentially catastrophic. If you are alarmed by the prospect of the worst-case scenarios sketched above, and yet also feel dissatisfied with the probable outcomes of the path that we are currently on, then as *individuals* we are called upon to be midwives of a more genial future. Undoubtedly, the fate of this fragile planet will largely be determined by the actions of powerful human institutions at local, national, and international levels. But we should not abdicate responsibility for radical system change to anonymously remote levels of the global hierarchy: to the CEOs, the directors general, the oligarchs, the phalanxes of ashen bureaucrats, the autocrats, the blue ribbon commissions, or even to our supposed technomessiahs. A groundswell of creativity, an unswerving commitment to justice, a comradely esprit, and a revolutionary zeal will be needed to implode moribund structures that trap us in despair. Those same capacities and inclinations are undoubtedly necessary to cultivate and sustain positive change.

References

1 Canetti, E. (1984). *Crowds and Power*. New York: Farrar, Straus and Giroux.
2 Meadows, D.H., Randers, J., and Behrens, W.W. (1972). *The Limits to Growth: A Report for the Club of Rome's Project on the Predicament of Mankind*, 27. Universe Books.
3 United Nations (2019). *World Population Prospects 2019: Highlights*. United Nations, Department of Economic and Social Affairs, Population Division.
4 Cohen, J.E. (1996). *How Many People Can the Earth Support?* WW Norton & Company.
5 Pörtner, H.-O. et al. (ed.) (2022). Climate change 2022: impacts, adaptation, and vulnerability. In: *Contribution of Working Group II to the Sixth Assessment Report of the Intergovernmental Panel on Climate Change (IPCC)*. Cambridge University Press.
6 Stanley, S.K. and Williamson, J. (2021). Attitudes towards climate change aid and climate refugees in New Zealand: an exploration of policy support and ideological barriers. *Environmental Politics* 30 (7): 1259–1280.
7 Riahi, K. et al. (2011). RCP 8.5 – a scenario of comparatively high greenhouse gas emissions. *Climatic Change* 109 (1, 2): 33–57.
8 Tainter, J. (1988). *The Collapse of Complex Societies*. Cambridge University Press.
9 Goldblatt, C. and Watson, A.J. (2012). The runaway greenhouse: implications for future climate change, geoengineering and planetary atmospheres. *Philosophical Transactions of the Royal Society A: Mathematical, Physical and Engineering Sciences* 370 (1974): 4197–4216.
10 Hansen, J. (2010). *Storms of My Grandchildren: The Truth About the Coming Climate Catastrophe and Our Last Chance to Save Humanity*. Bloomsbury Publishing.
11 Ashwin, P. and von der Heydt, A.S. (2020). Extreme sensitivity and climate tipping points. *Journal of Statistical Physics* 179 (5, 6): 1531–1552.

12 Franzke, C.L. (2014). Nonlinear climate change. *Nature Climate Change* 4 (6): 423–424.

13 Chipperfield, M.P. et al. (2015). Quantifying the ozone and ultraviolet benefits already achieved by the Montreal protocol. *Nature Communications* 6 (1): 1–8.

14 Solomon, S. et al. (1986). On the depletion of Antarctic ozone. *Nature* 321 (6072): 755–758.

15 Spratt, D. and Dunlop, I. (2019). *Existential Climate-Related Security Risk. A Scenario Approach.* National Centre for Climate Restoration (Breakthrough).

16 Franzen, J. (2019). What if we stopped pretending. *New Yorker* (8 September)

17 Wilson, E.O. (2016). *Half-Earth: Our Planet's Fight for Life.* WW Norton & Company.

18 Brondizio, E.S., et al., eds. Global assessment report on biodiversity and ecosystem services of the Intergovernmental Science-Policy Platform on Biodiversity and Ecosystem Services (IPBES). 2019. IPBES secretariat, Bonn, Germany.

19 Smil, V. (1999). Detonator of the population explosion. *Nature* 400 (6743): 415.

20 Smil, V. (2019). *Growth: from Microorganisms to Megacities.* MIT Press.

21 Kolbert, E. (2014). *The Sixth Extinction: An Unnatural History.* A&C Black.

22 Powell, T.W. and Lenton, T.M. (2013). Scenarios for future biodiversity loss due to multiple drivers reveal conflict between mitigating climate change and preserving biodiversity. *Environmental Research Letters* 8 (2): 025024.

23 Ritchie, H., Roser, M., and Rosado, P. (2020). *CO_2 and Greenhouse Gas Emissions.* Our World in Data.

24 Pindyck, R.S. (2011). Fat tails, thin tails, and climate change policy. *Review of Environmental Economics and Policy* 5 (2): 258–274.

25 Wagner, G. and Weitzman, M.L. (2016). *Climate Shock.* Princeton University Press.

26 Brysse, K. et al. (2013). Climate change prediction: erring on the side of least drama? *Global Environmental Change* 23 (1): 327–337.

27 Edwards, A.R. (2005). *The Sustainability Revolution: Portrait of a Paradigm Shift.* New Society Publishers.

28 Jemielniak, D. and Aibar, E. (2016). Bridging the gap between wikipedia and academia. *Journal of the Association for Information Science and Technology* 67 (7): 1773–1776.

29 Bateman, A. and Logan, D.W. (2010). Time to underpin Wikipedia wisdom. *Nature* 468 (7325): 765.

30 List of revolutions and rebellions. https://en.wikipedia.org/wiki/List_of_revolutions_and_rebellions (accessed December 2021).

31 Marcuse, H. (1972). Ecology and Revolution. In: *HerbertMarcuse, The New Left and the 1960s, Volume Three, Collected Papers of Herbert Marcuse* (ed. D. Kellner). London and New York: Routledge.

32 *Oxford English Dictionary.* Revolution, n. Oxford University Press (accessed December 2021).

33 Childe, V.G. (1950). The urban revolution. *The Town Planning Review* 21 (1): 3–17.

34 United Nations (2019). *The Future is Now – Science for Achieving Sustainable Development. Global Sustainable Development Report 2019.* New York: United Nations.

35 United Nations (2015). Transforming our world: the 2030 agenda for sustainable development. https://sdgs.un.org/2030agenda.

36 O'Neill, R.V. et al. (1986). *A Hierarchical Concept of Ecosystems.* Princeton University Press.

37 Rootes, C. (2004). Environmental movements. In: *The Blackwell Companion to Social Movements*, 608–640. Wiley.

38 Rootes, C. (2014). *Environmental Movements: Local, National and Global.* Routledge.

Index

A Primer on Human Impacts on the Environment: The Conceptual Approach, First Edition. Liam Heneghan.
© 2023 John Wiley & Sons Ltd. Published 2023 by John Wiley & Sons Ltd.